AF783265

GeoPlanet: Earth and Planetary Sciences

Editor-in-chief

Paweł Rowiński

Series editors

Marek Banaszkiewicz, Warsaw, Poland
Janusz Pempkowiak, Sopot, Poland
Marek Lewandowski, Warsaw, Poland
Marek Sarna, Warsaw, Poland

More information about this series at http://www.springer.com/series/8821

Vladimir Semenov · Maxim Petrishchev

Induction Soundings of the Earth's Mantle

Springer

Vladimir Semenov
Institute of Geophysics
Polish Academy of Sciences
Warsaw
Poland

Maxim Petrishchev
Pushkov Institute of Terrestrial Magnetism, Ionosphere and Radio Wave Propagation, Russian Academy of Sciences, St. Petersburg Branch
St. Petersburg
Russia

The GeoPlanet: Earth and Planetary Sciences Book Series is in part a continuation of Monographic Volumes of Publications of the Institute of Geophysics, Polish Academy of Sciences, the journal published since 1962 (http://pub.igf.edu.pl/index.php).

ISSN 2190-5193 ISSN 2190-5207 (electronic)
GeoPlanet: Earth and Planetary Sciences
ISBN 978-3-319-53794-8 ISBN 978-3-319-53795-5 (eBook)
DOI 10.1007/978-3-319-53795-5

Library of Congress Control Number: 2017946644

© Springer International Publishing AG 2018
This work is subject to copyright. All rights are reserved by the Publisher, whether the whole or part of the material is concerned, specifically the rights of translation, reprinting, reuse of illustrations, recitation, broadcasting, reproduction on microfilms or in any other physical way, and transmission or information storage and retrieval, electronic adaptation, computer software, or by similar or dissimilar methodology now known or hereafter developed.
The use of general descriptive names, registered names, trademarks, service marks, etc. in this publication does not imply, even in the absence of a specific statement, that such names are exempt from the relevant protective laws and regulations and therefore free for general use.
The publisher, the authors and the editors are safe to assume that the advice and information in this book are believed to be true and accurate at the date of publication. Neither the publisher nor the authors or the editors give a warranty, express or implied, with respect to the material contained herein or for any errors or omissions that may have been made. The publisher remains neutral with regard to jurisdictional claims in published maps and institutional affiliations.

Printed on acid-free paper

This Springer imprint is published by Springer Nature
The registered company is Springer International Publishing AG
The registered company address is: Gewerbestrasse 11, 6330 Cham, Switzerland

Series Editors

Geophysics

Paweł Rowiński
Editor-in-Chief
Institute of Geophysics
Polish Academy of Sciences
ul. Ks. Janusza 64
01-452 Warszawa, Poland
p.rowinski@igf.edu.pl

Space Sciences

Marek Banaszkiewicz
Space Research Centre
Polish Academy of Sciences
ul. Bartycka 18A
00-716 Warszawa, Poland

Oceanology

Janusz Pempkowiak
Institute of Oceanology
Polish Academy of Sciences
Powstańców Warszawy 55
81-712 Sopot, Poland

Geology

Marek Lewandowski
Institute of Geological Sciences
Polish Academy of Sciences
ul. Twarda 51/55
00-818 Warszawa, Poland

Astronomy

Marek Sarna
Nicolaus Copernicus Astronomical Centre
Polish Academy of Sciences
ul. Bartycka 18
00-716 Warszawa, Poland
sarna@camk.edu.pl

Managing Editor

Anna Dziembowska
Institute of Geophysics, Polish Academy of Sciences

Advisory Board

Robert Anczkiewicz
Research Centre in Kraków
Institute of Geological Sciences
Kraków, Poland

Aleksander Brzeziński
Space Research Centre
Polish Academy of Sciences
Warszawa, Poland

Javier Cuadros
Department of Mineralogy
Natural History Museum
London, UK

Jerzy Dera
Institute of Oceanology
Polish Academy of Sciences
Sopot, Poland

Evgeni Fedorovich
School of Meteorology
University of Oklahoma
Norman, USA

Wolfgang Franke
Geologisch-Paläntologisches Institut
Johann Wolfgang Goethe-Universität
Frankfurt/Main, Germany

Bertrand Fritz
Ecole et Observatoire des
Sciences de la Terre,
Laboratoire d'Hydrologie
et de Géochimie de Strasbourg
Université de Strasbourg et CNRS
Strasbourg, France

Truls Johannessen
Geophysical Institute
University of Bergen
Bergen, Norway

Michael A. Kaminski
Department of Earth Sciences
University College London
London, UK

Andrzej Kijko
Aon Benfield
Natural Hazards Research Centre
University of Pretoria
Pretoria, South Africa

Francois Leblanc
Laboratoire Atmospheres, Milieux
Observations Spatiales, CNRS/IPSL
Paris, France

Kon-Kee Liu
Institute of Hydrological
and Oceanic Sciences
National Central University Jhongli
Jhongli, Taiwan

Teresa Madeyska
Research Centre in Warsaw
Institute of Geological Sciences
Warszawa, Poland

Stanisław Massel
Institute of Oceanology
Polish Academy of Sciences
Sopot, Poland

Antonio Meloni
Instituto Nazionale di Geofisica
Rome, Italy

Evangelos Papathanassiou
Hellenic Centre for Marine Research
Anavissos, Greece

Kaja Pietsch
AGH University of Science and
Technology
Kraków, Poland

Dušan Plašienka
Prírodovedecká fakulta, UK
Univerzita Komenského
Bratislava, Slovakia

Barbara Popielawska
Space Research Centre
Polish Academy of Sciences
Warszawa, Poland

Tilman Spohn
Deutsches Zentrum für Luftund
Raumfahrt in der Helmholtz
Gemeinschaft
Institut für Planetenforschung
Berlin, Germany

Krzysztof Stasiewicz
Swedish Institute of Space Physics
Uppsala, Sweden

Ewa Szuszkiewicz
Department of Astronomy
and Astrophysics
University of Szczecin
Szczecin, Poland

Roman Teisseyre
Department of Theoretical Geophysics
Institute of Geophysics
Polish Academy of Sciences
Warszawa, Poland

Jacek Tronczynski
Laboratory of Biogeochemistry
of Organic Contaminants
IFREMER DCN_BE
Nantes, France

Steve Wallis
School of the Built Environment
Heriot-Watt University
Riccarton, Edinburgh
Scotland, UK

Wacław M. Zuberek
Department of Applied Geology
University of Silesia
Sosnowiec, Poland

Piotr Życki
Nicolaus Copernicus Astronomical
Centre
Polish Academy of Sciences
Warszawa, Poland

Theory should not be mistakenly
considered as reality which it describes

Stanislav Grof (1988)

Foreword

Interest in the Earth's geoelectrical structure and its time variability involves a strict requirement for the reliability of geophysical information due to the lack of direct examination methods like drillings in near-surface soundings. It concerns the crust and mantle induction soundings, which are available for investigation up to depths of about 1/3 of the Earth's radius ($\approx$2200 km). These soundings provide information about the Earth's interior by measuring natural electromagnetic field variations which constantly exist on the Earth. The advantage of the method is the fact that it is safe for the environment. However, its disadvantage is the complexity of the investigated signals from different field sources in the period range from minutes to months, related to the deep geoelectrical structure and even their time variability.

The basic idea of induction sounding is based on impedances. Their values are expected to be independent of time variations of the source power if external sources have known configuration of currents. These currents are generated by the Sun activity in the conductive ionosphere or magnetosphere of the Earth. However, the measured fields may include variations generated by other sources, like atmospheric processes on the Earth's surface, tidal waves, motions of ocean water streams in the Earth's magnetic fields, seismo-electric effects in sediments, biological fields in seas, or artificial currents having a high intensity now. All of them are noises for induction soundings. Therefore, the role of the theoretical basis of the deep induction soundings becomes extremely important. These requirements far exceed the approaches in exploration geophysics. Besides, long period impedances depend on types of field sources, among which there is already no place for a simplest "plane-wave" model. Indeed, the electromagnetic wavelengths on the monthly variations could be already comparable to the Earth's radius. It means that the main inhomogeneity can already be the Earth itself in non-conductive space. Thus, the Earth sphericity and its rotation in the primary fields should also be taken into consideration.

Those and other complicated problems of mantle soundings are analyzed by the authors in the book. Finally, the authors show interesting and unexpected regional time variabilities of secular Earth's apparent resistivity connected with Earthquake Quantity, and correlated with k- and aa-indexes and Wolf numbers.

Prof. Anatoly Guglielmi
Institute of Physics of the Earth Russian
Academy of Sciences, Moscow, Russia

Contents

Introduction

In this book, we present a generalized theoretical approach to the study of geo-electrical structure of the Earth's mantle, at the background of regional and global induction sounding results in Europe and Asia.

There are two types of deep induction soundings: magnetovariation (MVS) and magnetotelluric (MTS). Inasmuch as the Earth does not have its own impedances, they were postulated in different ways by various investigators. The aim of using impedances is to exclude time variations of source fields, keeping the geometry unchanged. Their space derivatives are based on the physical Induction Law in the MVS method, while the tippers are postulated separately in the MTS method. The forward modeling is based on physical laws. This 1D inversion is valid in the frame of laterally homogeneous, layered media with a possibility of recalculating the results for the spherical Earth. Such inversion can also include a possibility of taking into consideration the magnetic permeability of the layers.

The main complexity of the deep induction soundings is due to a great number of electromagnetic fields which cannot be considered as remote sources for deep soundings. The induction soundings are based on the idea of external fields generated by processes in the Earth's ionosphere and magnetosphere. Registered fields may be due to very diverse reasons: the processes in the lower atmosphere, the dynamics of the water masses in the ocean, and other processes in the lithosphere or oceans. First of all, these seismo-electromagnetic effects appear in some weak tectonic zones. The role of these factors increases for research in the low frequency band, i.e., in the secular deep soundings, for example.

The measured magnetic field is not the magnetic intensity **H** in A/m. SI units required by the IAGA in the Resolution 3 (of the year 1973) are used in this book. The magnetic induction **B** is measured in teslas (10^{-9} tesla is equal to 1 γ, which is an off-system unit). Thus, the induction impedances are $Z(\omega) = \pm E_i/B_j$ determined in m/s ($\check{Z} = E_i/H_j$ was an engineering impedance in ohms). So the field **B** should be used in the relations to find impedances and their derivatives or "we have to accept a measured response function $\mu \cdot \check{Z}$ instead of impedance Z" (Chave and Jones [eds.], 2012), where μ is the magnetic permeability. Note that "vectors **H** and **B** do not necessarily have the same directions if magnetic properties of media are anisotropic" (Atabekov [ed.], 1966). Impedances and admittances (impedance reciprocals) will be considered as scalars or matrices. The last ones can be transformed to tensors of apparent resistivity on the Earth's surface.

The authors are grateful to Profs. Leonid Vanyan, Anatoly Guglielmi, Ulrich Schmucker, Vladimir Shuman, Yuri Kopytenko, Pitter Weidelt, Mark Berdichevsky, Adam Schultz, Dr. Josef Pek, Assoc. Prof. Waldemar Jóźwiak, and Prof. Jerzy Jankowski for their fruitful, long cooperation. We also thank all our colleagues, including Anna Dziembowska, Dr. Anne Neska, Dr. Boris Ladanivsky, and Dr. Alexey Tkachev, for their remarks and help.

References

Atabekov, G.I. (ed.): Theoretical Bases of Electrical Engineering. Parts 2 and 3, p. 319. Energies, Moscow (1966) (in Russian)

Chave, A.D., Jones, A.G. (eds.): The Magnetotelluric Method: Theory and Practice, pp. 552. Cambridge University Press (2012)

Chapter 1
Impedances, Sources and Environments

Abstract The Earth does not have its own impedance and that is why the impedance definitions are numerous and, in some sense, arbitrary. Impedances adapt to the properties of the investigated media, introducing a priori information, for example, about their dimensionality. In this way, considering 1D, 2D or 3D environments means to mimic homogeneous and inhomogeneous media. Additionally, the geometry of external sources must be taken into consideration. As far as we know, the notion of "response function" appeared for the first time in the middle of the XIX century in the writings of J. Lamont [see in: Haak in The experiments with telluric currents and magnetic fields of Johann von Lamont in 1861, the sedimentary layer beneath Munich and the color theory of Johann Wolfgang von Goethe. Abstracts of 22 Electromagnetic Induction Workshops, (2014)] in Germany. At the end of the same century, Schuster and Lamb (Philosophical Transactions of the Royal Society of London. A 180:467–518, 1889) in the U.K. attempted to estimate the deep structure of the Earth using diurnal variations in Earths magnetic field for this purpose. Further development of the use of impedance occurs in the 1930s–1940s, probably due to the development of aircraft design, radar and active mineral exploration. The use of deep magnetovariation soundings is usually referred to the second half of the twentieth century, although sounding theory was developed as early as 1940 by S.M. Rytov and independently by A.N. Schukin in Russia.

Keywords Impedance · Source · Environment

1.1 Magnetotelluric Impedances in the XIX–XXI Centuries

The known response function, $C(\omega, r) = E/i\omega B$ (symbols explained below), was postulated more than 150 years ago, in 1861, by J. Lamont (see in: Haak 2014). It is independent of temporal variations of the source power but sensitive to changing geometry and position of source currents relative to the point of observation on the

© Springer International Publishing AG 2018

V. Semenov and M. Petrishchev, *Induction Soundings of the Earth's Mantle*,
GeoPlanet: Earth and Planetary Sciences, DOI 10.1007/978-3-319-53795-5_1

Earth's surface. Exception is the simplest "plane wave" model in which the vertical magnetic component must be absent. Since this would mean the absence of inhomogeneity in the Earth, an empirical relation with tippers has been added independently to fix the inhomogeneity in a medium separately. Magnetotelluric (MT) soundings are not based on physical laws, while Faraday's Law underlies IBC for evaluating magnetovariation impedances with their spatial derivatives.

Perhaps the definition of response function, $C(\omega, r) = E/i\omega B$, was first proposed by J. Lamont in 1861 (see in: Haak 2014) where horizontal electrical E and orthogonal magnetic B fields are written in the frequency domain, i is the imaginary unit, $\omega = 2\pi/T$ is the angular frequency and T is period. Later on, the impedances were defined by Hirayama (1934) and approximately at the same time by Leontovich (see in: Rytov 1940a).

Those impedances are still used in the magnetotelluric sounding (MTS) method, being known as Tikhonov (1950) and Cagniard (1953) model which is written below in the vector form (Landau and Lifshitz 1960):

$$E_\tau = Z(\omega) \cdot (B_\tau \text{ x n}) \tag{1.1}$$

Here $E_x(\omega) = Z(\omega) \cdot B_y(\omega)$ are complex Fourier amplitudes of the tangential electric and orthogonal magnetic fields, respectively, $Z(\omega)$ [m/s] is the magnetotelluric impedance, n is a unit vector normal to the surface, and τ is the index of lateral vectors. As a source for MTS method, the plane wave model is commonly used, in which the external component B_z is absent. Magnetotelluric impedances were derived for laterally homogeneous, one-dimensional (1D) isotropic media, in which a layered structure is possible. Results of induction sounding are traditionally presented as the apparent resistivity $\rho(\omega)$ for a layered isotropic media:

$$\rho(\omega) = [Z(\omega)]^2 \mu / i\omega \tag{1.2}$$

Here μ [H/m] is the magnetic permeability. This resistivity would not be "apparent" on a homogeneous conductive half-space.

However, practice required to estimate anomalous zones in layered structures of the Earth. That is why an additional empirical relation with dimensionless tippers (A and D) was suggested by Wiese (see in: Schmucker 1970 or Vozoff 1972) to describe a subsurface inhomogeneity independently of the impedance values:

$$A(\omega, \mathrm{r}) \cdot B_x + D(\omega, \mathrm{r}) \cdot B_y = B_z(\omega, \mathrm{r}) \tag{1.3}$$

Here r is the radius-vector from the fixed point (r = 0) up to points of observation. Note that $B_z(\omega, r)$ is the magnetic field component of internal origin only. So Eq. (1.3) is not related to MT soundings (1.1) which have no vertical component of the external magnetic field. In spite of this, the empirical relation (1.3) was successfully applied in practice to describe inhomogeneities in the shape of the following induction arrows:

$$\begin{aligned} \mathrm{Cu}(\omega, \mathrm{r}) &= \mathrm{Re}A \cdot \mathrm{e_x} + \mathrm{Re}D \cdot \mathrm{e_y} \\ \mathrm{Cv}(\omega, \mathrm{r}) &= \mathrm{Im}A \cdot \mathrm{e_x} + \mathrm{Im}D \cdot \mathrm{e_y}. \end{aligned} \tag{1.4}$$

Here e_x and e_y are vectors fixing two orthogonal directions of the corresponding values of tippers. Usually, induction arrows *Cu* are directed from a conductive body in ambient resistive media. Relation (1.4) with tippers helps fixing the conductive inhomogeneity. Above its center the vertical magnetic component is equal to zero, like above laterally homogeneous media.

The impedance boundary condition (IBC) for MT soundings was obtained by Rytov (1940a, b), Wait (1954) (see in: Senior and Volakis 1995). The MTS method can be applied for deep soundings on an ocean bottom, where the electric field noises are minimal at long periods.

1.2 Magnetovariation Impedances in the XX Century

Probably a first attempt to estimate the geoelectrical structure of the Earth's mantle was made by Schuster and Lamb (1889) using harmonics of daily oscillations. Since that time, several scientists were interested in this fascinating problem (for review, see Parkinson 1983). The daily oscillations of the magnetic field, variations during geomagnetic storms and (rarely) the polar electro-jets are used as source fields. Unfortunately, the resolving power attainable by the global method was not sufficient to reliably establish geoelectrical properties of the mantle (for review, see Rokityansky 1982). Later, taking into consideration non-shifted phases of impedances, the mid-mantle conductive zone, predicted earlier as a global structure by Zharkov (1983), was detected (Zakharova 1989; Schultz and Semenov 1993; and others). This zone, at depths from about 700–900 km, may have a conductivity comparably to that of sea water (Kelbert et al. 2009).

The magnetovariation sounding (MVS) method was developed in the XX century too. It could not be based on a plane-wave-like model because the vertical magnetic components of the sources have significant values in external source fields in all the above-mentioned cases. See, for example, the polar electrojets shown in Fig. 1.1. Moreover, these soundings are based on IBC following from two of Maxwell's equations: $rot\ E = -B/\partial t$ and $divB = 0$.

Experimental studies have already established that commonly the vertical component of induced electric field E_z in the Earth is by about four orders of magnitude less than the horizontal ones (Lipskaya 1971). But in some cases the amplitudes of E_z and E_τ can be comparable with each other (Jones and Geldart 1967a, b). It can be shown that $E_z = 0$ is justified in problems with horizontally homogeneous structure (Lipskaya 1971). So these commonly unmeasured values of spatial derivatives of electrical field components $\partial E_x/\partial z$, $\partial E_z/\partial x$, $\partial E_z/\partial y$ and $\partial E_y/\partial z$

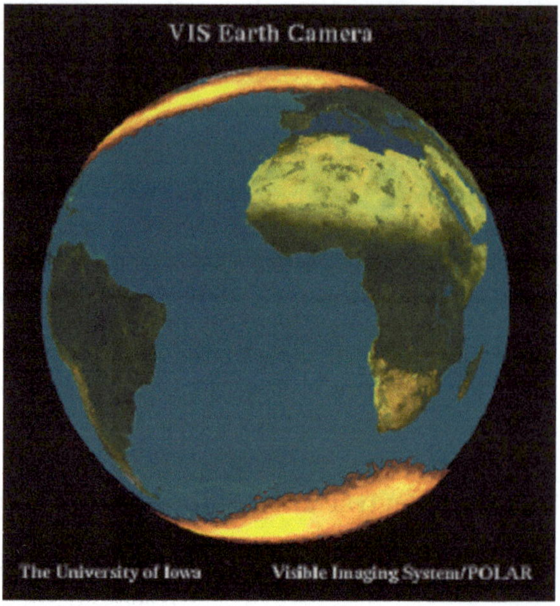

Fig. 1.1 Auroral currents (electrojets) in the ionosphere of the Earth. *Image Credit* the University of Iowa

have to be excluded from the analysis. That is why we write down here only the equation for vertical components of Faraday's Law in the frequency domain:

$$\partial E_x(\omega, \mathrm{r})/\partial \mathrm{y} - \partial E_y(\omega, \mathrm{r})/\partial \mathrm{x} \approx \mathrm{i}\omega B_\mathrm{z}(\omega, \mathrm{r}) \tag{1.5}$$

This equation can be applied to ensure the validity of Faraday's Law in experimental data and forward modeling.

If the Earth' medium is assumed to be laterally homogeneous, impedance derivatives are neglected and electric field components would be replaced by $E(\omega) = \zeta(\omega) \cdot B$ according to relation (1.1), then the IBC with magnetovariation impedance $\zeta(\omega)$ follows directly from relation (1.5):

$$\zeta(\omega) \cdot divB_\tau \approx i\omega B_z \tag{1.6}$$

Such a plane model was presented by Berdichevsky et al. (1969), Schmucker (1970) and Kuckes (1973). It allows finding impedances of laterally homogeneous, layered media from the IBC (Eq. 1.6) in the frame of a 1D model. Note that $divB_\tau(\omega)$ is equal to $-\partial B_z/\partial z$, since $divB = 0$. Impedances in relation (1.6) may be sensitive to relative directions of the point of measurement: they can be slightly changing during magnetic storms. Sounding on one magnetic storm is possible, but if it has a "sudden Dst beginning", the Laplace transformation is needed (Gokhberg 1966).

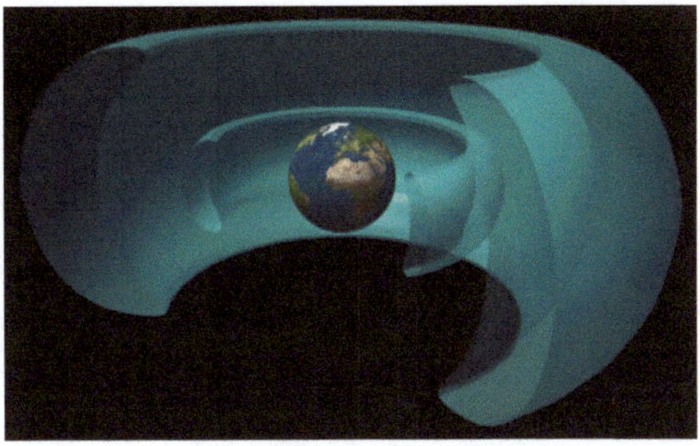

Fig. 1.2 Two ring currents in the Earth's magnetosphere. *Image Credit* NASA

The exciting magnetic fields for periods longer than a couple of days are caused by geomagnetic storms: the so-called Dst variations in the magnetosphere. They are approximately linearly polarized (Fuji and Schultz 2002) and induced by the ring currents along the geomagnetic longitudes (see Fig. 1.2). Then relation (1.6) can be expressed for the pure first zonal harmonic P_0^1 (Olsen 1998) on the spherical Earth in the corresponding form of the geomagnetic depth sounding (GDS) method as follows:

$$\zeta(\omega) \cdot 2B_\theta/R \cdot tg\theta = i\omega B_r \tag{1.7}$$

Here R is the Earth's radius, and $tg\theta$ is the geomagnetic co-latitude of the observation site. Exactly this expression, which is a spherical case of relation (1.6), was suggested by Banks (1969). Commonly, one sounding includes a time interval with several magnetic storms together.

Harmonics of daily (Sq) variations (1, ½, ¼ of day) are rarely used for deep magnetovariation and magnetotelluric soundings (Logvinov 2002; Ladanivskyy et al. 2010). They are too deep for practical purposes and rather complicated because the Earth's rotation must be taken into consideration.

For studies of the upper mantle impedances, the MTS and MVS methods can be combined. Impedances of both methods allowed to extend the analyzed period range and consequently to increase the depth of soundings up to about one third of the Earth's radius (e.g., Egbert and Booker 1992; Schultz et al. 1993; Schultz and Semenov 1993; Bahr et al. 1993). The appropriateness of this combination was based exclusively on prolongation of the MTS impedance modules and their phases by the MVS ones inside their confidence limits. Note that the phase data are more reliable than modules of MTS impedances because the latter can be shifted by surface inhomogeneity.

A magnetometer located on the Earth's surface scans the inhomogeneous magnetic field known as daily *Sq* time variations.

There are, generally speaking, two huge ring currents existing permanently and situated in the magnetosphere at a distance of 1.5–2 and 4–5 Earth's radii (Fig. 1.3)

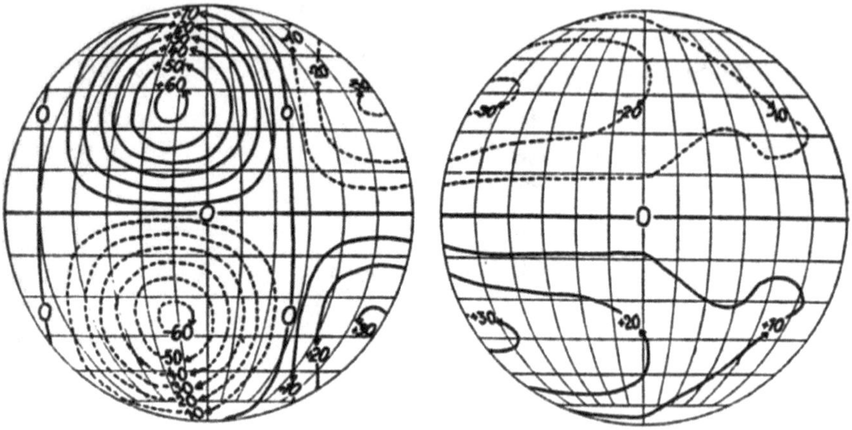

Fig. 1.3 Time-stable currents in the Earth's ionosphere due to its heating by the Sun at daily (*left*) and nightly (*center*) sides. After Janovski (1953)

with total intensities of 80,000 A and 1,100,000 A, respectively (Nishida 1978). The inner ring is fully controlled by the stable Earth's field and its own field can be neglected in comparison with the field of the outer ring current. The outer ring current is connected with the Sun-Earth interactions. It has an opposite direction than the inner one (Nishida 1978) and can be strongly amplified by the solar wind, causing the Dst variations or "magnetic storms". As a plausible model of the quiet ring current we consider the distant current belt (4.5 of the Earth radius) with width of about 200 km and center in a plane of the geomagnetic equator. The symmetry axis of this belt is considered to be also inclined by an angle of 11° to the Earth rotation axis. Note that for the model calculations we consider the eastward direction of the steady electric current in the distant belt, while in the real one it is directed westward.

Calculations show that the magnetospheric source producesa total field with an intensity $B_n \approx 24$ nT that is stable around the Earth and is directed parallel to geomagnetic longitudes. Using the theory presented above, the field components were calculated for the coordinates ($\vartheta = 35°$, $\lambda = 20°$) on the Earth's surface, as shown in Fig. 1.4. The part of this field, which is stable in time, is characterized by the spherical harmonics P_1^0 (cosϑ).

It is important to note that the MVS method has a depth limitation for soundings on the spherical Earth in contrast to the plane wave model for MTS. It is connected with wavelengths of source fields which can reach the Earth radius R at the long periods. The reasonable MVS sounding depth is about 1/3 of the Earth radius. It is equal to about 2000 km, that allows us to investigate the upper, middle, and top of the lower mantle.

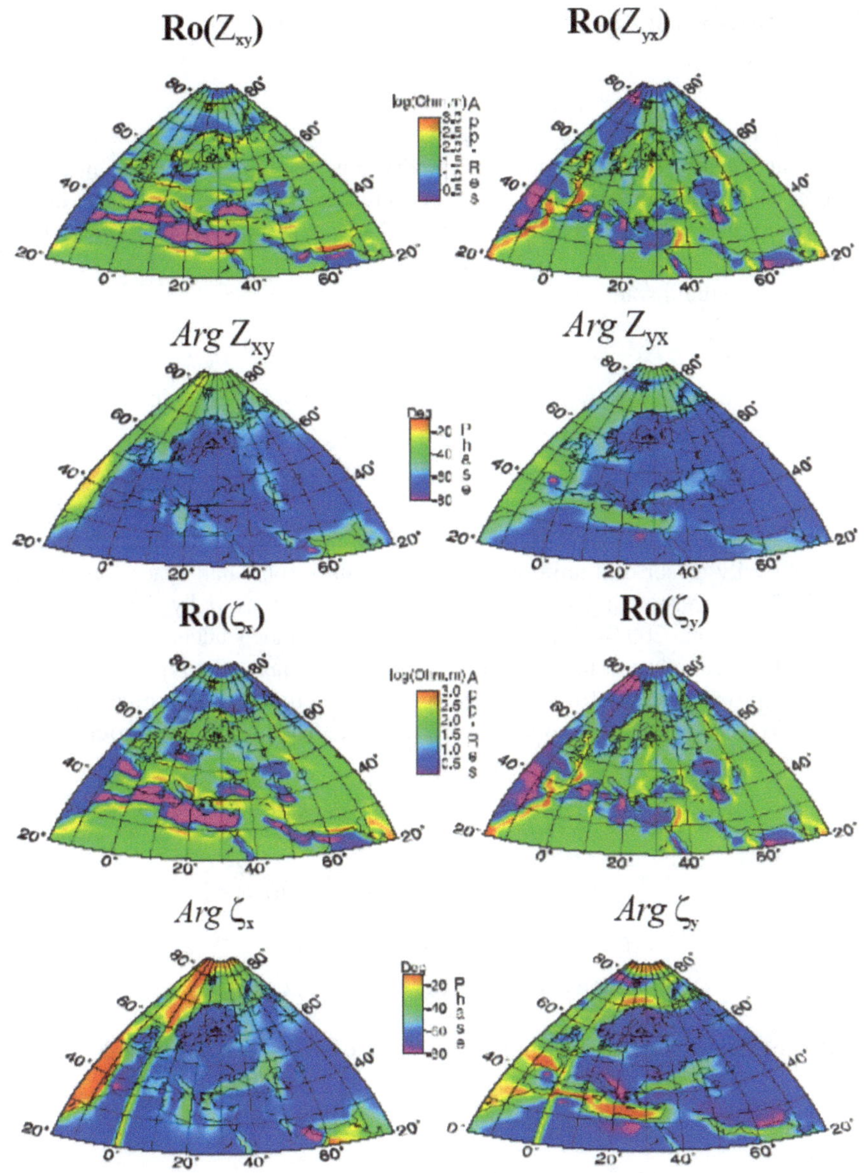

Fig. 1.4 Comparison of modeled apparent resistivities obtained for the northern hemisphere of the Earth with inhomogeneous subsurface conductivity. After Semenov and Shuman (2010)

1.3 Impedance Matrix for a Laterally Anisotropic Medium

Thus, for a homogeneous and isotropic medium the true resistivity can be estimated immediately. For a homogeneous but laterally anisotropic medium, the impedance $Z(\omega, \varphi)$ and admittances $Y_{ij}(\omega, \varphi)$ are considered as measured 2D matrices in two orthogonal directions (Landau and Lifshitz 1960; Senior and Volakis 1995). They were first applied by Berdichevsky and Cantwell (see in: Berdichevsky 1968) for the magnetotelluric sounding:

$$\begin{aligned}
E_x(\omega, \phi) &= Z_{xx}(\omega, \phi)\cdot B_x(\omega, \phi) + Z_{xy}(\omega, \phi)\cdot B_y(\omega, \phi)\\
E_y(\omega, \phi) &= Z_{yx}(\omega, \phi)\cdot B_x(\omega, \phi) + Z_{yy}(\omega, \phi)\cdot B_y(\omega, \phi)\\
B_x(\omega, \phi) &= Y_{xx}(\omega, \phi)\cdot E_x(\omega, \phi) + Y_{xy}(\omega, \phi)\cdot E_y(\omega, \phi)\\
B_y(\omega, \phi) &= Y_{yx}(\omega, \phi)\cdot E_x(\omega, \phi) + Y_{yy}(\omega, \phi)\cdot E_y(\omega, \phi)
\end{aligned} \tag{1.8}$$

Here φ is the azimuth, and $Y(\omega, \varphi)$ is the admittance equal to $Z(\omega, \varphi)^{-1}$. Thus, a true resistivity tensor of a uniform, azimuthally anisotropic half-space can also be found. The corresponding transformations were first derived by Reilly (see in: Weckmann et al. 2003). Then they were obtained independently by Semenov (1988, 2000) directly substituting the Eq. (1.8) in two Maxwell's Eq. (1.5). Here all fields are proportional to $\{-\exp t\}$. The spatial derivatives of field components are zero ($\partial/\partial x = \partial/\partial y = 0$) in a homogeneous medium. Then, as follows from the first equation above, these expressions are:

$$\begin{aligned}
\rho_{xx}(\omega, \phi) &= \left(Z_{xy}^2 - Z_{xx}\cdot Z_{yy}\right)\cdot \mu/i\omega\\
\rho_{yy}(\omega, \phi) &= \left(Z_{yx}^2 - Z_{xx}\cdot Z_{yy}\right)\cdot \mu/i\omega\\
\rho_{xy}(\omega, \phi) &= Z_{xx}\left(Z_{yx} - Z_{xy}\right)\cdot \mu/i\omega\\
\rho_{yx}(\omega, \phi) &= Z_{yy}\left(Z_{xy} - Z_{yx}\right)\cdot \mu/i\omega
\end{aligned} \tag{1.9}$$

Analogous expressions for admittances are the following (Semenov 2000):

$$\begin{aligned}
\rho_{xx}(\omega, \phi) &= \left(Y_{xy}^2 - Y_{xx}\cdot Y_{yy}\right)/Q^2\cdot \mu/i\omega\\
\rho_{yy}(\omega, \phi) &= \left(Y_{yx}^2 - Y_{xx}\cdot Y_{yy}\right)/Q^2\cdot \mu/i\omega\\
\rho_{xy}(\omega, \phi) &= Y_{yy}\cdot\left(Y_{xy} - Y_{yx}\right)/Q^2\cdot \mu/i\omega\\
\rho_{yx}(\omega, \phi) &= Y_{xx}\cdot\left(Y_{yx} - Y_{xy}\right)/Q^2\cdot \mu/i\omega
\end{aligned} \tag{1.10}$$

Here $Q(\omega) = \{Y_{yx}\cdot Y_{yx} - Y_{xx}\cdot Y_{yy}\}$ is the determinant of the admittance which is invariant under rotation of the coordinate system. Relations (1.9) and (1.10) show that the elements of impedance matrix, Z_{xy} and Z_{yx} both are zero, correspond to main apparent resistivities in the tensor of apparent resistivity, ρ_{xx} and ρ_{yy}, for isotropic media. And some inconsistence (see in: Adam 1966) disappears:

$$\begin{aligned} \rho_{x'x'} &= \rho_{xx} \cdot \cos^2\phi + \rho_{yy} \cdot \sin^2\phi - (\rho_{yx} + \rho_{xy}) \cdot \sin\phi \cdot \cos\phi \\ \rho_{x'y'} &= \rho_{xy} \cdot \cos^2\phi - \rho_{yx} \cdot \sin^2\phi + (\rho_{xx} - \rho_{yy}) \cdot \sin\phi \cdot \cos\phi \end{aligned} \tag{1.11}$$

These polar diagrams are useful for investigation of anisotropic layered media, a case studied in a series of publications, for example: Pek (2002), Pek and Santos (2002), Yin (2003). A careful study of the anisotropic media with the arbitrarily directed tensor of conductivity can be found therein, too. If a layered, azimuthally anisotropic medium is considered, these resistivities are complex, frequency-dependent quantities and they can be used as apparent resistivity tensors, too. Note that if the impedances Z_{xx} and Z_{yy} are zero, the diagonal elements of the resistivity tensor are connected with the off-diagonal elements of the impedance matrix.

Thus, for horizontally layered media there are two types of transformation to convert scalar impedances in isotropic media in any direction or impedance matrix in anisotropic media. The results of both transformations will be equal only if the directions in which one of the additional impedances or admittances points (see Eqs. 1.9 and 1.10) is equal to zero. These directions have been named by Professor Zhdanov "allotted directions", as opposed to "the major one" with two additional zero-valued impedances in perpendicular directions only. A special case is the so-called 2D model. In other arbitrary directions, the result of conversion will depend on the choice of transformation, i.e., isotropic or anisotropic model.

Apparent resistivities obtained from the two methods may be different. Calculations on test data have shown that the largest differences occur at subsurface depths and they can amount to more than 30% at mantle depths (Semenov 1998).

Note that conversion results of the measured impedances to the apparent resistivities according to Eqs. 1.3 or 1.5 will be different. The two conversions can produce curves of apparent resistivities with different shapes above inhomogeneous media.

For this reason, also their 1D inversion results can differ a few times in conductance values (Semenov 1998). The results of the two conversions will be the same only for directions at which both "principal directions" or one of two minor "preferential directions" of impedances are zero, i.e. $Z_{xx} \cdot Z_{yy} = 0$. This remark is important for combining the MTS "tensor" and the scalar MVS apparent resistivities or impedances.

The plane model has a narrow domain of applicability if for no other reason than undetermined ways of the current circuit closed at infinity because no boundary conditions exist in lateral directions. The assumption that the induced currents do not have back currents through the resistive crust is used for the subsurface soundings of the layered media. Such a model can satisfy demands of the exploration activity and can be successful for this aim. However, this model is usually incorrect for the mantle soundings because of the absence of "plane waves" in the spherical model. Besides, the presence of subduction and spreading zones and deep faults allow the circuits to be closed also through the crust.

1.4 3D Impedance Matrix

A more general theory with six impedance values, where all electric and magnetic field components in mutually orthogonal directions are included, was considered by Dmitriev and Berdichevsky (2002) for MT soundings beneath an exciting vertical field. The authors have assumed that only the first derivatives of the exciting field can exist. In this case, they had to consider the MT impedance as a 3×3 matrix instead of a 2×2 one (1.4) or a scalar (1.7), i.e., the linear relationships between Fourier amplitudes of all field components can be considered in the form suggested by Berdichevsky and Zhdanov (1984) in the Cartesian co-ordinates:

$$\begin{aligned}
E_x &= Z_{xx} \cdot B_x + Z_{xy} \cdot B_y + Z_{xz} \cdot B_z \\
E_y &= Z_{yx} \cdot B_x + Z_{yy} \cdot B_y + Z_{yz} \cdot B_z \\
E_z &= Z_{zx} \cdot B_x + Z_{zy} \cdot B_y + Z_{zz} \cdot B_z \\
B_x &= Y_{xx} \cdot E_x + Y_{xy} \cdot E_y + Y_{xz} \cdot E_z \\
B_y &= Y_{yx} \cdot E_x + Y_{yy} \cdot E_y + Y_{yz} \cdot E_z \\
B_z &= Y_{zx} \cdot E_x + Y_{zy} \cdot E_y + Y_{zz} \cdot E_z
\end{aligned} \quad (1.12)$$

Because the corresponding IBCs are unknown, we assume that $E_z = 0$ on the Earth's surface and inside the isotropic medium for a pure TE induction mode (the current circuits are closed in horizontal directions) and $B_z = 0$ for the pure TM galvanic mode (the circuits are closed in vertical plane). Replacement of the relations with the 3×3 impedance matrix (1.9) by relations (1.4) with the 2×2 matrix has been analyzed. Then the relationships of admittances for the pure TE mode (1.4) remain, while those with the impedance (1.4) are transformed for this mode as follows:

$$\begin{aligned}
E_x^{TE} &= \left\{Z_{xx}^{TE} + a \cdot Z_{xz}^{TE}\right\} \cdot B_x^{TE} + \left\{Z_{xy}^{TE} + b \cdot Z_{xz}^{TE}\right\} \cdot B_y^{TE} \\
E_y^{TE} &= \left\{Z_{yx}^{TE} + a \cdot Z_{yz}^{TE}\right\} \cdot B_x^{TE} + \left\{Z_{yy}^{TE} + b \cdot Z_{yz}^{TE}\right\} \cdot B_y^{TE} \\
B_x^{TE} &= Y_{xx}^{TE} \cdot E_x^{TE} + Y_{xy}^{TE} \cdot E_y^{TE} \\
B_y^{TE} &= Y_{yx}^{TE} \cdot E_x^{TE} + Y_{yy}^{TE} \cdot E_y^{TE}
\end{aligned} \quad (1.13)$$

Here $a(\omega, r) = -Z_{zx}^{TE}/Z_{zz}^{TE}$ and $b(\omega, r) = -Z_{zy}^{TE}/Z_{zz}^{TE}$. The analogous expressions for the pure TM-mode can be easily rewritten:

$$\begin{aligned} E_x^{TM} &= Z_{xx}^{TM} \cdot B_x^{TM} + Z_{xy}^{TM} \cdot B_y^{TM} \\ E_y^{TM} &= Z_{yx}^{TM} \cdot B_x^{TM} + Z_{yy}^{TM} \cdot B_y^{TM} \\ B_x^{TM} &= \{Y_{xx}^{TM} + c \cdot Y_{xz}^{TM}\} \cdot E_x^{TM} + \left\{Y_{xy}^{TM} + d \cdot Y_{xz}^{TM}\right\} \cdot E_y^{TM} \\ B_x^{TM} &= \left\{Y_{yx}^{TM} + c \cdot Y_{yz}^{TM}\right\} \cdot E_x^{TM} + \left\{Y_{yy}^{TM} + d \cdot Y_{yz}^{TM}\right\} \cdot E_y^{TM} \end{aligned} \tag{1.14}$$

Here the transfer functions are $c(\omega, r) = -Y_{zx}^{TM}/Y_{zz}^{TM}$ and $d(\omega,r) = -Y_{zy}^{TM}/Y_{zz}^{TM}$. The relationships for impedances and admittances (1.10 and 1.11) are unsymmetrical for each mode. Thus, impedances for the pure TM mode (1.11) as well as admittances for the pure TE mode (1.12) are equal to the ones estimated from relationships (1.4).

However, both modes have to be combined for the mantle soundings of inhomogeneous media. Under obvious assumptions $E_\tau = E_\tau^{TE} + E_\tau^{TM}$, $B_\tau = B_\tau^{TE} + B_\tau^{TM}$, $E_z = E_z^{TM}$, $B_z = B_z^{TE}$ and taking into account that $B_x^{TM} = a \cdot B_x^{TE}$, $B_y^{TM} = b \cdot H_y^{TE}$ for the time-harmonic signals of the same source, the relationships (1.10) and (1.11) in the presence of both modes can be combined:

$$\begin{aligned} E_x &= \{[Z_{xx}^{TE} + a \cdot Z_{xz}^{TE}] + b \cdot Z_{xx}^{TM}\} \cdot B_x^{TE} + \left\{\left[Z_{xy}^{TE} + c \cdot Z_{xz}^{TE}\right] + d \cdot Z_{xy}^{TM}\right\} \cdot B_y^{TE} \\ E_y &= \left\{\left[Z_{yx}^{TE} + a \cdot Z_{yz}^{TE}\right] + b \cdot Z_{yx}^{TM}\right\} \cdot B_x^{TE} + \left\{\left[Z_{yy}^{TE} + c \cdot Z_{yz}^{TE}\right] + d \cdot Z_{yy}^{TM}\right\} \cdot B_y^{TE} \end{aligned}$$

In a brief form, these expressions have been presented by Becken and Pedersen (2000):

$$E_\tau^{TE} + E_\tau^{TE} = \left(Z^{TE} + Z^{TM}\right) \cdot B_\tau^{TE}$$

These relations show the complexity of data analysis of this kind, because the measurements in these two modes are not made separately and simultaneously. Strictly speaking, according to the theorem of Harrington (1961), all currents will be closed within a closed surface only. The plane model has no closed surface.

1.5 Impedances for Laterally Inhomogeneous Media

Following the adage "Everything new is actually well-forgotten old" let us return back to the year 1940: Rytov has published his article in French (Rytov 1940a) and in Russian (Rytov 1940b). This work has remained unnoticed for a long time except of the publication of Rytov's teacher (Leontovich 1948) and then (Guglielmi and Gokhberg 1987). Below, shorthand IBC relation between the Fourier amplitudes of magnetic field components is written out replacing the modern impedance $Z = E/B$. Now let us return to Faraday's Law written above (Eq. 1.5) for its vertical terms:

$$\text{rot}_z E_\tau(\text{t, r}) \approx -\partial B_Z(\text{t, r})/\partial \text{t} \tag{1.15}$$

The essential difference between relations (1.15) and (1.5) is the presence of the radius-vector r which indicates an additional dependence on the point of observation.

The Fourier amplitudes of both fields will be considered in the form: $B(\omega, \text{r}) \cdot \text{e}^{-i\omega t}$ and $E(\omega, \text{r}) \cdot \text{e}^{-i\omega t}$, where $\omega = 2\pi/T$ is the angular frequency, and T is the period in seconds. Thus the equality presented above can be rewritten in the frequency domain (see relation 1.6). This equation may be applied to ensure fulfillment of this component of Faraday's Law in experimental data and forward modeling.

Following Guglielmi and Gokhberg (1987), some of Rytov's IBCs for an isotropic, smoothly inhomogeneous medium can be written, for MTS and MVS methods, in approximate forms as follows:

$$-E_y \approx Z(\omega, \text{r}) \cdot \text{B}_\text{x} + (Z(\omega, \text{r})/2\text{k}) \cdot \left[\partial^2 B_\text{x}/\partial \text{x}^2 - \partial^2 B_\text{y}/\partial \text{y}^2 + 2 \cdot \partial^2 B_\text{y}/(\partial \text{x} \cdot \partial \text{y})\right]$$

$$-\text{i}\omega\,\text{Bz}(\omega, \text{r}) \approx Z(\omega, \text{r}) \cdot [\partial \text{Bx}/\partial \text{x} + \partial \text{By}/\partial \text{y}] + \text{B}_\text{x} \cdot \partial Z(\omega, \text{r})/\partial \text{x} + B_\text{y} \cdot \partial Z(\omega, \text{r})/\partial \text{y} \tag{1.16}$$

Here, the wave number $\text{k}^2 = -i\omega\mu\sigma$ and the transfer function $Z(\omega, \text{r}) = (i\omega\mu/\sigma)^{1/2}$ is the scalar impedance, where $\sigma(\omega, \text{r})$ is the specific effective conductivity of the conductive half-space.

Let us substitute the two orthogonal electric components expressed through impedance definitions $E_\text{x} = Z(\omega, \text{r}) \cdot B_\text{y}$, $E_\text{y} = -Z(\omega, \text{r}) \cdot B_\text{x}$ (Landau and Lifshitz 1960) in second relation (1.16). The spatial derivatives of impedance are dimensional quantities, 1/s, and the plane divergence of the magnetic field $\partial B_x/\partial x + \partial B_y/\partial y = -\partial B_z/\partial z$, since $divB = 0$. So the gradient of the vertical magnetic field may be included in this relation instead of divergence of magnetic field.

Note that the equation $Z \cdot (\partial B_x/\partial x + \partial B_y/\partial y) = i\omega B_z$ suggested for MVS method by Berdichevsky and Schmucker is a special case of Eq. (1.17) (see below). The only difference is that relation (1.6) has no terms with spatial derivatives of impedances. Besides, it was shown (Olsen 1998) that the relation $Z \cdot (2 \cdot B_\theta/R \cdot \text{tg}\theta_0) = i\omega B_z$ suggested by Banks (1969), follows directly from the Berdichevsky-Schmucker relation written in a spherical reference system (R is the Earth radius, θ_o is the co-latitude of the point of observation) for magnetospheric ring currents as source. So it is also a special case of IBC (1.14) in a spherical reference system.

Comparison of relation (1.3) with relation (1.14) shows that the term $i\omega$ was missing in the right part of (1.3) for the case div $B = 0$. That is why tippers $A(\omega, \text{r})$ and $D(\omega, \text{r})$ are dimensionless quantities and Eq. (1.3) must be considered as an empirical one, but not as a special case of Eq. (1.14). Although such an empirical approach is often sufficient in practice for subsurface soundings controlled by drillings, this approach is doubtful for deep soundings because the vertical magnetic field is a powerful part of external source fields causthe model were written ased, for example, by harmonics of Sq variations.

If the external field is assumed as a "plane wave" ($B_z = 0$) and the layered medium is plane, too, and isotropic, then all spatial derivatives of fields and impedances are zero: $\partial Z/\partial x = \partial Z/\partial y = \partial B_x/\partial x = \partial B_y/\partial y = \partial E_x/\partial y = \partial E_y/\partial x = 0$ and consequently: $Z \cdot 0 + 0 \cdot B_x + 0 \cdot B_y = i\omega B_z = 0$. Then IBC (1.14) is losing its meaning. However, the simplest impedances can be found at each measurement point according to their definitions presented above and can be inverted in the frame of a 1D model. The correlated 1D cross-section or depths of the total conductance beneath these points can be considered as a model for interpretation (e.g., Semenov and Jozwiak 2006). Local anomalies can be detected even without knowledge about impedance values due to occurrence of a vertical magnetic component.

Significant dispersions of the apparent resistivities at different observatories can occur if the Earth's sphericity is not taken into account (Semenov et al. 2011) for long distances between observatories. Such effect can be minimized if Eq. (1.2) is considered in the spherical reference system with impedance $Z_s(\omega, R, \varphi, \lambda)$:

$$\begin{aligned} &Z_s \cdot \left[\partial\left(B_\phi \cdot \sin\phi\right)/\partial\phi + \partial B_\theta/\partial\lambda\right]/(R \cdot \sin\phi) \\ &\quad + \left(B_\phi/R\right) \cdot \partial Z_s/\partial\phi + \left(B_\theta/R \cdot \sin\phi\right) \cdot \partial Z_s/\partial\lambda = i\omega B_r \end{aligned} \tag{1.17}$$

If the response function $C(\omega, r) = Z(\omega, r)/\omega$ is considered for deep soundings, then Eq. (1.14) must be rewritten as:

$$C \cdot \left(\partial B_x/\partial x + \partial B_y/\partial y\right) + (\partial C/\partial x) \cdot B_x + (\partial C/\partial y) \cdot B_y = iB_z$$

The spatial derivatives of the response function $C(\omega, r)$ are not a dimensional quantity like tippers and will be useful for analyzing the induction vectors.

This IBC is a generalized MVS method and an analogous relation exists for the generalized MTS method too (Senior and Volakis 1995) which is still not applied for the deep mantle soundings. These IBCs allow estimating impedance $Z = E/B$ together with their first spatial derivatives at the Earth's surface from experimental data. This impedance approach has started from IBC at a contact between resistive (air) and conductive (earth) media. Such relations between the Fourier amplitudes of the harmonics of field components follow from IBC derived by Wait (1954). By the way, the above-mentioned work has considered another possibility of performing Earth's soundings.

This sounding differs fundamentally from the cases considered above. The point is that if a conductivity heterogeneity exists in an environment, it usually generates gradients in the observed fields, which are corresponding to tippers but with another value.

The traditional dimensionless induction arrows *C*u & *C*v (Schmucker 1970) are two vectors based separately on real and imaginary parts of complex tippers *A* and *D* (1.3). These vectors are directed along two unit vectors, e_x and e_y, coinciding with orthogonal directions of measurements. They were written above for fixed r and ω (1.4). The induction *vectors* *S*u & *S*v are defined similarly from two complex and dimensional spatial derivatives of impedance in the same directions:

$$\begin{aligned} Su &= Re(\partial Z/\partial x) \cdot e_x + Re(\partial Z/\partial y) \cdot e_y \\ Sv &= Im(\partial Z/\partial x) \cdot e_x + Im(\partial Z/\partial y) \cdot e_y \end{aligned} \tag{1.18}$$

These *vectors* have to be distinguished from the traditional dimensionless induction *arrows* *C*u & *C*v (Schmucker 1970), because they were found from essentially different equations: empirical (1.3) or theoretical (1.2), respectively. The vectors *S*u & *S*v have a physical sense: they are gradients of real and imaginary parts of complex impedances. The induction vectors can be also rewritten using response function C(r, ω) = *Z*(r, ω)/ω:

$$Wu = Su/\omega; Wv = Sv/\omega. \tag{1.19}$$

So it is sufficient to divide magnitudes of vectors *S*u & *S*v by ω (their directions remain the same) and the new vectors *W*u and *W*v will be also dimensionless like tippers.

Note that the response function *C*(r, ω) = *Z*(r, ω)/ω and consequently vectors *S*u/ω and *S*v/ω differ from the definition of Schmucker: *C'*(r, ω) = *Z*(r, ω)/*i*ω; the imaginary unit *i* was ignored in the denominator which leads to swapped complex parts of these vectors. This may be the reason why the imaginary part of the response function *C'*(r, ω) was used by Professor Ulrich Schmucker to estimate skin depths.

Moreover, the forward spherical modeling of both kinds of induction impedances (1.1, 1.3, 1.4) has shown their mutual inconsistence above inhomogeneities in the mantle (Vozar and Semenov 2010). That is why, in practice, combinations of impedances obtained by both induction methods can meet problems hampering in this way the mantle investigations. In particular, this concerns the period range from some hours to a couple of days, and it is most difficult when induction soundings over several different sources of the exciting field are presented simultaneously (see in: Utada et al. 2008). So perhaps we need to consider more complicated IBCs to estimate impedances on an inhomogeneous medium.

Then, the IBCs were defined in a more accurate way, as well as in modified and generalized forms (e.g., Boerner and Ahluwalia 1972; Aboul-Atta and Boerner 1975; Bates et al. 1976; Senior 1981; Kartashov et al. 1985; Senior and Volakis 1995; Shuman 2003, 2007; Guglielmi 2009). In the following, we will shortly review the impedance types with the same physical origin and separate them to obtain possible combinations for the practical aim of mantle soundings.

A number of relations that are more general than (1.1)–(1.4) can be found in Rytov (1940a, b). These IBCs follow immediately from Maxwell's equations and tolerate that the exciting field is caused by a distant source (not a "plane wave"), that the surface is curved (not entirely flat), that smoothed lateral anomalies can exist (at a distance greater than the wavelength in the conductive media) and consequently, the impedance depends on coordinates, too. They were derived for radio-wave frequencies in form of a power series with a small parameter characterizing a skin-depth for a well-conducting medium. For example, following Guglielmi and Gokhberg (1987), some of Rytov's IBCs for an isotropic, smoothly inhomogeneous medium can be written in approximate form, in the Cartesian co-ordinates, as follows: the transfer function, $Z(\omega, \mathrm{r}) = (i\omega\mu/\sigma)^{1/2}$, is a scalar impedance, where $\sigma(\omega, \mathrm{r})$ is the specific effective conductivity of the conductive half-space. Assuming that field components as well as lateral characteristics of media are not changing at all along the surface, the first expression of (1.6) is the same as relation (1.1) (Leontovich 1948) and the second one transforms to $B_z(\omega, \mathrm{r}) = 0$. So, it is the standard "plane wave" model by Cagniard and Tikhonov for a laterally homogeneous half-space.

If we assume that spatial derivatives of the second and higher orders of the tangential magnetic field are negligible quantities, we obtain both MT and MV induction soundings in the following vectorial approximate forms (Guglielmi and Gokhberg 1987; Guglielmi 2009; Senior and Volakis 1995):

$$E_\tau \approx Z(\omega, \mathrm{r}).(B_\tau, \mathrm{n}) \tag{1.20}$$

$$Z(\omega, \mathrm{r})\mathrm{div}\, B_\tau + B_\tau \cdot \mathrm{grad}\, Z(\omega, \mathrm{r}) \approx \mathrm{i}\,\omega B_r \tag{1.21}$$

The last relationship has been named the generalized horizontal spatial gradient (gHSG) sounding to distinguish it from its special case, the horizontal spatial gradient (HSG) sounding, for the laterally homogeneous media (1.6). Note that impedance at both relationships is congruent as confirmed by the numerical modeling (Semenov et al. 2007). Relationship (1.8) has been applied in the implicit forms by Woods and Lilley (1979), Kuckes et al. (1985), and it has been considered as a "scattering" characteristic (Bates et al. 1976; Kuckes et al. 1985).

The gHSG method includes a relatively new term: $B_\tau \cdot gradZ(\omega, \mathrm{r})$, with the spatial derivatives of the impedance. In the plane wave model there applies $divB_\tau = 0$. So this IBC looks like the well-known Wiese-Parkinson expression, with the exception that the vertical component of the magnetic field is preceded by a factor $i\omega$:

$$B_\tau \cdot gradZ = (\partial Z/\partial \mathrm{x}) \cdot B_\mathrm{x} + (\partial Z/\partial \mathrm{y}) \cdot B_\mathrm{y} = i\omega B_z$$

where $\partial Z/\partial \mathrm{x}$ and $\partial Z/\partial \mathrm{y}$ are the gradients of the impedances. Now they can be also used to construct the induction arrows with a clear physical sense: they are spatial gradients of the real and imaginary parts of the scalar magnetovariation response function C. Note that here the vertical magnetic field component is caused by an internal source, the horizontal components of which can produce non-zero $div \cdot B_\tau$, which had been assumed as negligible. If the medium is isotropic and laterally homogeneous, i.e., *gradZ* = 0, relation (1.8) is naturally transformed to the HSG method (1.2).

A relationship similar to (1.8) has been obtained by Schmucker (2003) as the combination of the HSG sounding and the Wiese-Parkinson expression. Besides, the scalar impedance was extended by an impedance matrix. However, the considered IBC (1.8) has the advantage that it involves the observed field components everywhere. Therefore, a complicated iterative procedure for separating the fields into their "normal" and "anomalous" parts required by the former approach can be avoided.

These relationships can be used directly in the radio- and perhaps audio-induction soundings. However, the wavenumbers for the mantle sounding periods (i.e., from hours to years) are small in contrast to the radio-wave frequencies. Thus, the Rytov power series may not converge at all for these periods and corresponding effective conductivities of our planet. A similar situation exists for media with frequency dispersion of electrical characteristics (Kopeikin 1998). So, we have to continue our search for an approach or for more reliable substantiations for applying the relationships presented above to the mantle soundings.

1.6 Modern Impedances for Soundings

Another approach has been suggested by Aboul-Atta and Boerner (1975) which requires no conditions for inhomogeneity or limitation of period range. The authors made use of the theorem that the electromagnetic fields inside a medium can be found uniquely if the tangential fields are known on its *closed* surface (Harrington 1961). They have defined a new "vectoral impedance boundary condition" (VIBC) on the surface: two projections of complex vector $E_\tau(\omega, \mathrm{r})$ on complex vector $B_\tau(\omega, \mathrm{r})$ require two orthogonal bases in a 2D unitary vector space (Shuman and Kulik 2002) for a unique solution to exist. The VIBCs for the model were written as follows (Shuman 1999):

$$E_\tau = \zeta \cdot (B_\tau \times \mathrm{n}) + \xi^* \cdot B_\tau^*, \tag{1.22}$$

Here n is the unit vector of the outer normal to the interface and the asterisk (*) means the complex conjugate values. According to the unicity theorem, only two impedances, $\zeta(\omega, \mathrm{r})$ and $\xi(\omega, \mathrm{r})$, can satisfy this VIBC (Shuman 2007) in spherical coordinates:

$$\zeta_\alpha = \left(E_\theta B_\phi^* - E_\phi B_\theta^*\right)/||B_\tau||^2 \quad \text{and} \quad \xi_\alpha^* = \left(E_\theta B_\theta + E_\phi B_\phi\right)/||B_\tau||^2 \tag{1.23}$$

where $||B_\tau||^2 = B_\varphi^2 + B_\theta^2$ and the parameter α specifies a direction of the field polarization. These new impedances can be found if the complex Fourier amplitudes of the tangential electric and magnetic field components are known on a *full* spherical surface.

This approach has been tested by numerical simulations. The modeling has shown equivalence of the impedances Z (1.1) and ζ (1.8) above homogeneous isotropic media, for which ξ^* is equal to zero (Prichepiy 2006). The new and old approaches have been compared under the condition that both of them exist simultaneously: $E_\tau = \zeta \cdot (B_\tau\ n) + \xi^* \cdot B_\tau{}^* = Z_{ij} \cdot B_\tau$. It was established that they can produce different impedances already for the principal directions of a 2D model (Prichepiy 2007).

The comparison of both approaches was also carried out by the numerical simulation on the layered globe (Kuvshinov et al. 2005) with a surface conductance shell (Vozar et al. 2006). The results are presented in Fig. 1.4 for the European region in terms of apparent resistivities (1.2) and impedance phases. It is visible that the corresponding impedances are similar, but they also have differences, especially above the sharp inhomogeneity on the continental scale.

The new approach has been applied to the MV sounding by Shuman (1999):

$$-i\omega\mu \cdot B_{\rm n} = \zeta(divB_\tau) + B_\tau \cdot \left(grad\zeta\right) + \xi^*\left(divB_\tau^*\right) + B_\tau^* \cdot \left(grad\ \xi^*\right) \qquad (1.24)$$

The impedances $\zeta(\omega, \mathrm{r})$ and $\xi^*(\omega, \mathrm{r})$ are the same as the MT ones (1.13). If $\xi^* \approx 0$, the last expression is congruent to relationship (8), but without its theoretical restrictions.

Additionally, two tangential vectors, $L(\omega, \mathrm{r})$ and $K(\omega, \mathrm{r})$, follow from the VIBC (1.12) (Shuman 2007):

$$L = \zeta(E^* \times B) - \xi^*(E \times B^*) = n \times K$$

They are orthogonal, purely imaginary, and they decrease faster with distance from the inhomogeneity than the common induction arrows do (Prichepiy 2007). They have to be normalized before using them in a similar way as the standard induction arrows.

The individual approaches considered above produce sets of specific impedances for the deep induction methods. They are equal to each other above laterally homogeneous, isotropic media. But already for azimuthally anisotropic or inhomogeneous ones, impedances obtained for the MT and MV soundings can be different. The impedances of the daily harmonics (their currents are changing directions during one day) will be effective and scalar, while those for long period Dst variations (their currents are polarized along the geomagnetic longitude) will reflect the property of the medium in this particular direction only. Thus, impedances are dependent on the sounding methods, the assumed media properties, the accepted space models without or with the relative motions of the conductive media, and on the exciting fields with their source field structure. This confirms the statement that "the Earth does not have its own impedance" (Guglielmi and Gokhberg 1987). Thus, combinations of impedances obtained by different methods must be substantiated theoretically or by forward modeling, for example.

The approach for subsurface soundings of laterally homogeneous media excited by a "plane wave" source as well as investigation of conductive anomalies by using the induction arrows can satisfy requirements of the exploration activity. However, it may be risky to apply the approach directly to the soundings of the Earth's mantle where both, TE and TM modes can co-exist simultaneously in the same direction (Becken et al. 2008) due to inhomogeneities of the subduction type and spreading zones, which allow the currents to penetrate through the crust. Besides, the spherical model is preferred for the mantle investigations on the regional scales, at least to avoid uncertainty with current circuits closed at infinity in the plane model.

The new VIBC approach, including Rytov's approximate relationships as a special case, appears appropriate as the theoretical basis for the deep induction soundings. It does not require any conditions limiting inhomogeneity, sources, or wave numbers, except for the closed surface. This approach is promising in terms of forward modeling (Vozar and Semenov 2010) and testing in practice of the deep soundings (Schmucker 2003). It has a "rather high immunity to near-surface galvanic distortions" (Berdichevsky et al. 2006).

Sometimes it seems that the problem of combining impedances obtained by different methods could be avoided by rejection of impedances at all. In fact, the modern numerical simulations allow carrying out a direct comparison of the modeled and observed fields both for periodical signals (Olsen and Kuvshinov 2004) and for transient Dst fields in the time domain (Velimský and Martinec 2005). Moreover, real source structures, with signals modeled by a random process, are applied randomly (Vanyan et al. 2002). Thus, the concept of impedances must not be overemphasized. A specific combination of the MT and MV methods for mantle induction soundings has been applied for regional investigations by Sokolova and Varentsov (2007) and by Semenov et al. (2008).

1.7 Influence of Non-linear Ohm's Law

According to experimental data, the ores, particularly sulphide minerals, can be characterized by a non-linear conductivity (Shaub 1965, 1971; Izmailov and Silantev 1973, 1976). Impressive experimental results were obtained by Izmailov and Silantev (1976), showing a non-linear relationship between the current and the electrical field for several kinds of minerals like those presented in Fig. 1.5. This effect can depend on frequency (Sheynman 1969), pressure and/or temperature (Shuy 1979). Of course, such effects may be detected in nature only if the magnetotelluric variations have a very high intensity and the anomalously conductive ores are wide-spread. At any rate, the possibility of investigating this type of electrical conductivity of the Earth's crust or mantle from experimental data is interesting enough to estimate this effect theoretically.

Let us consider Ohm's law in a more general, non-linear form instead of the one usually used in magnetotelluric theory. For some direction, for example, the expression can be written as follows:

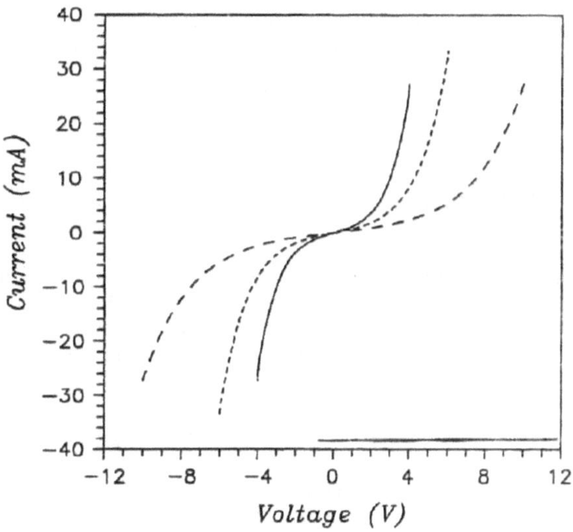

Fig. 1.5 Approximation of experimental data obtained by Izmailov and Silantev (1973) for sulphide minerals. Different lines mean different non-linear conductivity of medium

$$j = \sigma \cdot E + \sigma \cdot \Sigma \lambda_n \cdot E^n, \quad n = 1, 2, 3....$$

Here j is the current density, $E(\omega)$ is the electric field, and λ_n are coefficients describing curves of Volt–Ampere characteristics for some ores. Details are in Semenov (1998) but the main idea is that signal can exist at all multiple frequencies simultaneously, while the source field consists only one. In other words, the process of the energy spreading through the frequency range can appear if we are dealing with the non-linear conductivity in the Earth. The correlation between these signals at different frequencies contains information about the non-linear conductivity of medium.

1.8 Conclusion

In this chapter we have presented the history of attempts to use the variations of electromagnetic fields for study the Earth's interior. Despite the fact that the earth does not have its own impedance, this concept is widely used in the theory and practice of electromagnetic soundings. There are described the impedance boundary conditions and sources for different methods of soundings including the modern and difficult case—the vectoral impedance boundary conditions.

It should be remembered that "Everything new is actually well-forgotten old". We have used the strong theory obtained by Rytov and published in French (1940a) and Russian (1940b). It was corrected later in a special book written by Senior and Volakis (1995). This theory was independently obtained by Schukin (1940).

Particular attention is paid to the dimension of the medium. In addition, the chapter presents the results of interesting studies of the non-linearity of Ohm's law, which is not reflected in the sounding theory.

References

Aboul-Atta, O.A., Boerner, W.M.: Vectoral impedance identity for the natural dependence of harmonic fields on closed boundaries. Can. J. Phys. **53**(15), 1404–1407 (1975)

Adam, A.: Uber den Vergleich der elektromagnetischen Componenten von Observatorien Nagycenk and Tihany an ruhigen tagen (Sq) auf Grund der Daten vom Jahre 1961. Geofizikai Kozlemenyek **XVII**(1–2), 39–50 (1966)

Bahr, K., Olsen, N., Shankland, T.J.: On the combination of the magnetotelluric and the geomagnetic depth sounding method for resolving an electrical conductivity increase at 400 km depth. Geophys. Res. Lett. **20**(24), 2937–2940 (1993)

Banks, R.J.: Geomagnetic variations and the electrical conductivity of the upper mantle. Geophys. J. R. Astron. Soc. **17**, 457–487 (1969)

Bates, R.H.T., Boerner, W.M., Dunlop, G.R.: An extended Rytov approximation and its significance for remote sensing and inverse scattering. Opt. Commun. **18**(4), 421–423 (1976)

Becken, M., Pedersen, L.B.: Transformation of VLF anomaly maps into apparent resistivity and phase. Geophysics **68**(2), 497–505 (2000)

Becken, M., Ritter, O., Burkhart, H.: Mode separation of magnetotelluric responses in the three-dimensional environments. Geophys. J. Int. **172**, 67–86 (2008)

Berdichevsky, M.N.: Electrical Prospecting by the Method of Magnetotelluric Profiling. Nedra, Moscow (1968) (in Russian)

Berdichevsky, M.N., Vanyan, L.L., Fainberg, E.B.: Frequency sounding of the earth by spherical analysis of electromagnetic variations. Geomag. Aeron. **9**, 372–374 (1969) (in Russian)

Berdichevsky, M.N., Zhdanov, M.S.: Advanced Theory of Deep Geomagnetic Sounding. Elsevier, Amsterdam (1984)

Berdichevsky, M.N., Dmitriev, V.I., Pushkarev, P., Golubtsova, N., Kuznetsov, V.: A defence of the magnetovariational method. IAGA WG 1.2 on Electromagnetic Induction in the Earth, Extended Abstract 18-th Workshop, El Vendrell, Spain, September 17–23 (2006)

Boerner, W.M., Ahluwalia, H.P.S.: On a set of condition wave electromagnetic inverse boundary condition. Can. J. Phys. **50**(23), 3023–3061 (1972)

Cagniard, L.: Basic theory of the magneto-telluric method of geophysical prospecting. Geophysics **18**(3), 605–635 (1953)

Dmitriev, V.I., Berdichevsky, M.N.: A generalized model of impedance. Izvestiya Phys. Solid Earth **38**(10), 897–903 (2002)

Egbert, G.D., Booker, J.R.: Very long period magnetotellurics at Tucson observatory: implications for mantle conductivity. J. Geophys. Res. **97**(B11), 5099–15112 (1992)

Fujii, I., Schultz, A.: The 3D electromagnetic response of the Earth to ring current and aurorally oval excitation. Geophys. J. Int. **151**(3), 689–709 (2002). doi:10.1046/j.1365-246X.2002.01775.x

Gokhberg, M.B.: The possibilities of using geomagnetic storms to study the electrical characteristics of the earth. Dokl. Acad. Sci. USSR **166**(4), 851–853 (1966) (in Russian)

Guglielmi, A.V.: On the Leontovich boundary condition in geoelectromagnetism. Izvestiya Phys. Solid Earth **45**(9), 740–743 (2009). doi:10.1134/S106935130909002X

Guglielmi, A.V., Gokhberg, M.B.: On magnetotelluric sounding in seismically active regions. Fizika Zemli **33**(11), 122–123 (1987). (in Russin)

Haak, V.: The experiments with telluric currents and magnetic fields of Johann von Lamont in 1861, the sedimentary layer beneath Munich and the color theory of Johann Wolfgang von Goethe. Abstracts of 22 Electromagnetic Induction Workshops (2014)

Harrington, R.F.: Time harmonic electromagnetic fields, p. 480. McGraw-Hill, New York (1961)

Hirayama, M.: On the relations between the variations of Earth potential gradient and terrestrial magnetism. J. Meteorol. Soc. Jpn. Ser. II **12**(1), 16–22 (1934) (in Japanese with English abstract)

Ismaylov, L.I., Silantyev, V.N.: Non-linear electrical properties of sulfide ores. Transaction of Far East Department of the Academy of Sciences of the USSR. New geological data of the Northeast of USSR. 55, Magadan, pp. 230–235 (1973) (in Russian)

Ismaylov, L.I., Silantyev, V.N.: Non-linear Voltage-Ampere characteristics of sulfide ores. Transaction of Far East Department of the Academy of Sciences of the USSR. Geophysical investigation of the Earth crust. 76, Magadan, pp. 56–67 (1976) (in Russian)

Janovski, B.M.: Earth's Magnetism. 2nd edition (1953) (in Russian)

Jones, F.W., Geldart L.P.: Vertical telluric currents. Earth Planet. Sci. Lett. **2**(1), 67–74 (1967a)

Jones, F.W., Geldart, L.P.: Vertical telluric currents at separated locations. Earth Planet. Sci .Lett. **2**(3) (1967b)

Kartashov, YuA, Parantaev, G.V., Yachno, YuL: On the approximate boundary conditions on the quasi-flat conductive layer in the geomagnetic and geoelectrical tasks. Izvestiya Phys. Solid Earth **31**(9), 55–60 (1985) (in Russian)

Kelbert, A., Adam, Schultz A., Egbert, G.: Global electromagnetic induction constraints on transition-zone water content variations. Nature **460**, 1003–1007 (2009)

Kopeikin, V.: Description of the dispersive refraction effect by space-time ray method. Wave Motion **27**, 307–319 (1998)

Kuckes, A.F., Nekut, A.G., Thompson, B.G.: A geomagnetic scattering theory for evaluation of the Earth structure. Geophys. J. R. Astron. Soc. **8**, 319–330 (1985)

Kuvshinov, A., Utada, H., Avdeev, D., Koyama, T.: 3D modeling and analysis of the Dst EM responses in the North Pacific Ocean REGION. Geophys. J. Int. **160**, 505–526 (2005)

Ladanivskyy, B.T., Semenov, VYu., Logvinov, I.M.: Magnetovariation Sounding Method of Earth's Mantle at Period Band 104–105 sec. (Ukrainian). Geophys. J. **32**(3), 50–59 (2010) (in Russian)

Landau, L.D., Lifshitz, E.M.: Electrodynamics of Continuous Media. Pergamon Press, Oxford (1960)

Leontovich, M.A.: On approximate boundary conditions for an electromagnetic field on the surface of highly conductive bodies. In: Investigation of the radio waves propagation. The Academy of Sciences of the USSR, Moscow, 5–12 (1948) (in Russian)

Lipskaya, N.V.: Some patterns of distribution of inhomogeneous field in the horizontally homogeneous medium. The natural electromagnetic field and study the internal structure of the Earth. Publishing House "Nauka", Moscow, 7–15 (1971) (in Russian)

Logvinov, I.M.: Applying of the horizontal spatial gradient method for the deep conductivity estimations in the Ukraine. Acta Geophys. Pol. **50**(4), 567–573 (2002)

Nishida, A.: Geomagnetic Diagnostics of the Magnetosphere. Springer, Berlin (1978)

Olsen, N.: The electrical conductivity of the mantle beneath Europe derived from C-responses from 3 to 720 hr. Geophys. J. Int. **133**, 298–308 (1998). doi:10.1046/j.1365-246X.1998.00503.x

Olsen, N., Kuvshinov, A.: Modeling the ocean effect of geomagnetic storms. Earth Planet Space **56**, 525–530 (2004)

Parkinson, W.D.: Introduction to Geomagnetism. Scottish Academic Press, Edinburgh (1983)

Pek, J.: Spectral magnetotelluric impedances for an anisotropic layered conductor. Acta Geophys. Polonica. **50**(4), 619–643 (2002)

Pek, J., Santos, A.M.: Magnetotelluric impedances and parametric sensitivities for 1-D anisotropic layered media. Comput. Geosci. **28**, 939–950 (2002)

Prichepiy, T.I.: Modeling of the impedance parameters of the electromagnetic fields. Geophys. J. Kiev **28**(1), 121–129 (2006) (in Russian)

Prichepiy, T.I.: Vector identity of impedance type and ellipses of polarization of harmonic electromagnetic field. Geophys. J. Kiev **29**(5), 124–142 (2007) (in Russian)

Rokityansky, I.I.: Geoelectromagnetic investigation of the earth's crust and mantle. Springer, Berlin, New-York (1982)

Rytov, S.M.: Calcul du skin-effect parla methode des perturbations. Journal de Phisique USSR **2**, 233–242 (1940a) (in French)

Rytov, S.M.: Skin-effect calculations by the disturbance method. J. Exp. Theory Phys. **10**(2), 180–189 (1940b) (in Russian)

Schmucker, U.: Anomalies of geomagnetic variations in the southwestern United States. Bull. Scripnogrs. Inst. Ocean. **13**, 165 pp (1970)

Schmucker, U.: Horizontal spatial gradient sounding and geomagnetic depth sounding in the period range of daily variation. In: *Protokoll über das Kolloquium elektromagnetische Tiefenforschung*, Kolloquium: Königstein, pp. 228–237 (2003)

Schukin, A.N.: Propagation of Radiowaves. Svyazizdat, Moscow (1940) (in Russian)

Schultz, A., Kurtz, R.D., Chave, A.D., Jones, A.G.: Conductivity discontinuities in the Upper Mantle beneath a stable craton. Geophys. Res. Lett. **20**(24), 2941–2944 (1993)

Schultz, A., Semenov, V.Y.: Modelling of electro conductivity structure of mid-mantle of the Earth. Izvestia Russ. Acad. Sci. Earth Phys. **10**(25), 221–226 (1993) (in Russian)

Schuster, A., Lamb, H.: The diurnal variation of terrestrial magnetism. Philos. Trans. R. Soc. Lond. A **180**, 467–518 (1889)

Semenov, V.: Analysis of magnetotelluric data for sounding of an anisotropic media. Geol. Geophys. **10**, 121–125 (1988) (in Russian)

Semenov, V.: On the apparent resistivity in magnetotelluric sounding. Izvestiya Phys. Solid Earth **36**(1), 99–100 (2000)

Semenov, V., Jozwiak, W.: Lateral variations of the mid-mantle conductance beneath Europe. Tectonophysics **416**, 279–288 (2006)

Semenov, V., Ladanivskyy, B., Nowożyński, K.: New induction sounding tested in Central Europe. Acta Geophys. **59**(5), 815–832 (2011)

Semenov, V., Shuman, V.N.: Impedances for the deep electromagnetic soundings. Acta Geophys. **58**(4), 527–542 (2010)

Semenov, V., Vozar, J., Shuman, V.: A new approuch to gradient geomagnetic sounding. Izvestiya Phys. Solid Earth **43**(7), 592–596 (2007)

Semenov, V.: Regional conductivity structures of the Earth's mantle. Publication of Institute of Geophysics PAS, vol. C-65, issue no (302), 122 pp (1998)

Semenov, V., Ádám, A., Jóźwiak, W., Ladanyvskyy, B., Logvinov, I.M., Pek, J., Pushkarev, P., Voza, J.: Experimental team of CEMES, 2008. Electrical structure of the upper mantle beneath Central Europe: results of the CEMES project. Acta Geophysica **56**(4), 957–981 (2008)

Senior, T.B.A., Volakis, J.L.: Approximate Boundary Conditions in Electromagnetics, p. 353. IEE Press, London (1995)

Senior, T.B.A.: Approximate boundary conditions. IEEE Trans. Antennas Propag AP **29**(5), 826–829 (1981)

Shaub, Y.B.: On the usage of non-linear conductivity effect of ores at the electrical methods. Acad. Sc. USSR. Earth Phys. **6**, 76–81 (1965) (in Russian)

Shaub, Y.B.: Observations of non-linear effects from double electrical layer by aftograf method. J. Geoph. Apparatus. **46**, 26–31 (1971) (in Russian)

Sheynman, S.M.: Modern physical basis of electrical method theory. Nedra, Leningrad, 222 p. (1969) (in Russian)

Shuman, V.N.: Imaginary surface vectors in multidimensional inverse problems of geoelectrics. Izvestiya Phys. Solid Earth **43**(3), 592–596 (2007)

Shuman, V., Kulik, S.: The fundamental relations of impedance type in general theories of the electromagnetic induction studies. Acta Geophys. Polon. **50**(4), 607–618 (2002)

Shuman, V.: The general theory of geoelectromagnetic sounding systems accounting the electrodynamics of spherical sources. *3D EMIII Workshop*, Adelaide (2003)

Shuman V.N.: Scalar Local Impedance Conditions and the Impedance Tensor in Processing and Interpretation of a Magnetotelluric Experiment. Geophysical J., Kiev. 19, 361–385 (1999)

Shuy, R.T.: The Semiconductor Mineral Ores. (Nedra, Leningrad, 288 p. (1979) (in Russian)

Sokolova, EYu., Varentsov, I.M.: Deep array electromagnetic sounding on the Baltic Shield: external excitation model and implications for upper mantle conductivity studies. Tectonophysics **445**, 3–25 (2007)

Tikhonov, A.N.: On determining electrical characteristics of the deep layers of the earth's crust. Doklady AN. USSR. **73**(2), 259–297 (1950) (in Russian)

Utada, H., Yoneda, A., Shimizu, H., Baba, K., Palshin, N.A.: Effect of Sq variations on the electromagnetic response functions in the period range between 3 hours and 1 day. IAGA WG 1.2 on Electromagnetic Induction in the Earth, 19th Workshop, Beijing, China, October 23–29, p. 848 (2008)

Vanyan, L.L., Kuznetsov, V.A., Lyubetskaya, T.V., Palshin, N.A., Korja, T., Lahti, I., The BEAR Working Group.: Electrical conductivity of the crust beneath Central Lapland. Izv. Phys. Solid Earth **38**(10), 798–815 (2002)

Velimský, J., Martinec, Z.: Time domain harmonic-finite element approach to transient three-dimensional geomagnetic induction in a spherical heterogeneous earth. Geophys. J. Int. **161**, 81–101 (2005)

Vozoff, K.: The magnetotelluric method in the exploration of sedimentary basins. Geophysics **37**, 98–141 (1972)

Vozar, J., Semenov, V.: Compatibility of induction methods for mantle soundings. J. Geophys. Res. **115**, B03101 (2010). doi:10.1029/2009JB006390

Vozar, J., Semenov, V., Kuvshinov, A.V., Manoj, C.: Updating the map of Earth's surface conductance. Eos. Trans. AGU **87**(33), 326–331 (2006)

Wait, J.R.: On the relation between telluric currents and the Earth's magnetic field. Geophysics **19**, 281–289 (1954)

Weckmann, U., Ritter, O., Haak, V.: Images of the magnetotelluric apparent resistivity tensor. Geophys. J. Int. **155**(2), 456–468 (2003). doi:10.1046/j.1365-246X.2003.02062.x

Woods, D.V., Lilley, F.E.M.: Geomagnetic induction in Central Australia. J. Geomagn. Geoelect. **31**, 449–458 (1979)

Yin, C.: Inherent non-uniqueness in magnetotelluric inversion for 1D anisotropic models. Geophysics **68**(1), 138–146 (2003)

Zakharova, O.K.: About the data interpretation of global magnetovariation soundings. Geomag. Aeron. **29**(1), 167–169 (1989) (in Russian)

Zharkov, V.N.: Internal Structures of the Earth and Planets. Nauka, Moscow (1983) (in Russian)

Chapter 2
Several Impedances from One Equation

Abstract The chapter is dedicated to estimate impedances and their space derivatives using the theory of the random processes. This approach is effective to estimate several unknown values from one equation. Results are prepared in the frequency domain and characterized by confidence limits. Selections of directions for interpretation are discussed. The theoretical principles considered below deal with fields induced by sources in the ionosphere or magnetosphere of the Earth only.

Keywords Random processes · Spectra · Transfer function

2.1 Introduction

The purpose of data processing of induction soundings data is the definition of impedance—the main characteristics of soundings. The induction soundings are carried out on natural fields, the intensity and position of which vary with time. Impedances, representing a ratio of the spectra of the field components, are independent of the power source variation, at least at the time of observation. Impedance depends on the frequency, on the oscillations of external fields, electrical conductivity and magnetic permeability of the medium, which allows us to perform soundings. Note that impedances are also depending on the source characteristics of the field. So, following Guglielmi, we can say: "The Earth does not have its own impedance"—it's just a convenient feature for analysis.

Most of the magnetotelluric source fields are not stable and that is why the impedances determined at different times, i.e. for different combinations of the field sources, may be different from each other. Matching of all sources of the field is impossible, and there is a need to use the criteria for selecting the best impedance or a combination of them. Such criteria are either of negative character, as the impossibility of increasing the impedance value with the increase in the period of field variations, or based on a priori knowledge of the signal/noise ratio in individual components of the observed field.

© Springer International Publishing AG 2018

V. Semenov and M. Petrishchev, *Induction Soundings of the Earth's Mantle*,
GeoPlanet: Earth and Planetary Sciences, DOI 10.1007/978-3-319-53795-5_2

It is most fundamental to have a theoretical criterion for measuring the link between spectra of the observed field components. This is the coherence. Usage of this criterion on the basis of the existing theory of harmonic signals is not possible: a linear relationship between the Fourier amplitudes of sine waves on the same frequency is always there! Furthermore, one field component is generally simultaneously connected not to two, but to three or more other field components. We know from school that one equation with two or more unknowns has an infinite number of solutions. However, if the registered electromagnetic fields can be simulated by a random process with some limitations, it is possible to quantify the strength of linear relationships between their components in the frequency band. As a measure of linear relationship there appears the coherence—an analog of correlation in the frequency domain.

Due to complicated multi-variate connections between the components of the field we are forced to introduce more detailed criteria, like partial and multiple coherences. In addition, the representation of the field variations by a random process allows in many cases to establish criteria for selecting the best of the impedances found by means of a priori information about the nature of the field and how to measure it. The result of this analysis is the evaluation of the impedance within the frequency bands and the amplitudes with a predetermined probability.

As in any statistics, the fulfillment of such criteria is a necessary but not sufficient condition for the existence of impedances. Therefore, the analysis of data based on the theory of random processes only warns the researchers about the false conclusions; at best, it points to the need to improve the techniques of observation and theoretical foundations of the method. Checking of the validity of the selected criteria remains to be done by a comparison of results of interpretation of sounding with the independent research.

2.2 Modeling of Fields by Stochastic Processes

2.2.1 Random Process Definition

At the end of the last century, a lot of papers were dedicated to the problems of spectral analysis of geophysical data (Bath 1974; Reddy and Rankin 1974; Svetov and Shimilevich 1982; and so forth). However, even in these papers the random process was used not as a strict mathematical model but as an illustration of application of stochastic approaches to data analysis. In terms of the theory of random processes it is not difficult to justify the possibility of evaluating solutions of an equation with two or three unknowns in the range of frequencies and amplitudes with a given probability. In the approach presented below, the random process is considered as a model of registered fields for induction sounding of the Earth.

A field component variation is generally written as X (t), where t is time. We agree that the time interval of the registration changes in the infinite limits.

Processes in nature do not occur in isolation: they are accompanied by a number of factors not accounted for in the particular experiment. In other words, registered data are not reproducible and could be different, depending on unrecorded factors.

In this case, it is useful to use a set of functions, only one of which is our registered data and the remaining simulate all kinds of other outcomes of a similar experiment. That would be considered once the whole set of features is mathematically written as X (t, χ), where χ is a parameter called an elementary event, fixing a specific function of the proposed set. This set is a random process that we will use as a model of the observed field.

If we fix any value of the parameter $\chi = \chi_0$, then the process will be only a function of time $X = X(t, \chi_0)$ and this is the realization of a random process. A random processes is the set of all its realizations. If we fix any moment of time t_0, then it will be only a function of elementary events $X = X(t_0, \chi)$, which is called a random variable.

What gave such definitions? Let the components of the field entry X (t) be modeled by a random process $X(t, \chi)$. It means that the resulting data is the realization of a random process, and the value of the random variable is registered at each time. From these statements we know nothing about any other possible values of the field. The restrictions described below are imposed on the random process so that it could serve as a specific model of the electromagnetic field variations and to obtain reliable data in the framework of this model.

2.2.2 *The Distribution Function and Its Moments*

First of all, we should set the probability for a space of elementary events. To do this, it should be determined what are the acceptable values of electromagnetic field, which could be registered at time t_0? Let us formulate the answer as follows: it is possible to have any field values, but with different probabilities. Indeed, the full-scale data show that if we exclude the regular oscillations of the field from the consideration, then we can see that significant amplitude variations occur not as often as the variation of smaller amplitude. The probability to register a variation that is great in magnitude (greater than the field of the Earth) is practically zero in the time scales used in practice. This response should be formulated now as restrictions to the random process.

As it is stated that the random variable $X(t_0, \chi)$ may have any value, it is clear that the set of elements χ must be large enough. Further, the probability space should be measurable, which is equivalent to the existence of a random process. In this space we define a measure, hereinafter simply referred to as probability *P*: to each subset of the space we assign a number that is less than one, and to the set we assign one. It should be noted that such a construction of measure approximately corresponds to the strict definition. For example, we assume *P* to be an area, and the area of the entire set is equal to 1. In this case, we would get an even distribution, unacceptable for us. To set the probability we can determine all finite-dimensional distribution functions. To do this, we must assume that all implementations are continuous functions of time. Such a requirement is not contrary to the observed fields without instantaneous changes in their values.

Let us define: the finite-dimensional distribution function is a function that determines the probability of an elementary event in which the values of the random variables do not exceed the specified values. The adopted field model has an infinite number of implementations and we can safely say that it is continuous. It should also require its continuous differentiability, which will consider the moments of the distribution functions, such as expectation, variance, etc. Please note that the periodic components of the field, which is the base for soundings theory, were excluded from consideration. Otherwise, the distribution density function cannot satisfy the condition of its continuity (Bendat and Piersol 1986; Jenkins and Watts 1968).

The density distribution function can be fully characterized by all its moments, the first two of which are the expectation and variance. The expectation of a random process used below is defined as follows:

$$M[X(t, \chi_i)] = \int X(t, \chi)P(d\chi) \tag{2.1}$$

The indefinite integral of measure function P follows from the introduction of the space of elementary events. In practice, this estimation is performed using the integral of the elementary functions and, for example, in the case of a process with a finite number of realizations, takes the form:

$$M[X(t, \chi_n)] = \sum_{n=1}^{N} X(t, \chi_n)P(\chi_n). \tag{2.2}$$

The expectation, as well as all other aspects of the distribution function, is a non-random characteristic of a random process, and, as a result of integrating (2.1), it does not depend on parameter χ anymore.

The following example is important for understanding the mathematical expectation of a random process. The expectation of a random process consists of one implementation, for example a sine wave, and this coincides with the realization that follows from (2.2). Hence, the expectation of a sine wave is itself a sinusoid, and an intuitive understanding of its average value of zero corresponds to the expectation of a random process consisting of a plurality of sine waves: $X(t, \chi) = \cos[\omega t + \Psi(\chi)]$, where $\Psi(\chi)$ is uniformly distributed from $-\pi$ to π. Obviously, in general, the expectation of the random process is time-dependent, as well as the distribution function itself. Similarly, all its moments may have this property as well. This fact greatly complicates the practical application of our model to the electromagnetic field. In the random model, it is necessary to impose additional conditions.

2.2.3 *Stationary and Ergodic Hypothesis*

Of course, considering all the finite-function distributions is almost impossible. Thus, we assume that all of the distribution functions are invariant with respect to any changes in time. Processes which satisfy this condition are called strictly stationary. In the stationary random process the expectation does not depend on time. Next, we will consider only such processes and therefore the term "stationary" can be omitted.

The electromagnetic field seems to satisfy this condition only in a first approximation. The intensity of the field sources is naturally changing over time. However, considering the variability of the field is useful for a specific purpose only. This can be, for example, a marked influence of transients on the sounding results or the study of the variability of transfer functions in time. The theory of induction sounding was built to established processes and does not cover large time scales. Thus, as long as there are sufficient grounds for the introduction of a specific type of unsteadiness in a random field model, it is better not to enter. In addition, practical conclusions can be drawn by segments' implementations that in extreme cases can be adjusted to criteria of stationary processes within the registered segments studied. In such a way, it may be possible to study the variability of the spectra over time.

In general, a stationary random process can be characterized by a distribution function. The distribution law of a random process for modeling field variations is usually taken to be normal. The normal distribution law is simple, since it can be fully characterized by the value of the mathematical expectation and variance. The variance of a stationary process is independent of time and can be found through the autocorrelation function:

$$R_{xx}(\tau) = M[X(t,\chi)X(t+\tau,\chi)] \tag{2.3}$$

The value of autocorrelation function at $\tau = 0$ is the variance of a random process.

For a simultaneous study of two field components, each being a realization of a random process with its own characteristics, the cross-correlation function is used:

$$R_{xx}(\tau) = M[X(t,\chi)\,Y(t+\tau,\chi)] \tag{2.4}$$

Equation (2.4) holds for the consideration of a two-dimensional random process.

To determine the characteristics of the non-random process, one must know all its implementations, since the integration is expected in the whole space of elementary events. Such information is not available to the researcher. In addition to the original information, a random process includes all sorts of outcomes of a similar experiment, which is impossible to observe. They can be understood as the field variations recorded in a similar experiment, e.g. registered in the same place at another time.

Then, combining a large number of records at one point in time, it would be possible to evaluate the characteristics of the non-random process. This would be natural if the characteristics determined in such a way were close to their values defined for each implementation individually. These examples of arguments illustrate to some extent the introduction of the ergodic hypothesis: a stationary random process satisfies this hypothesis, if for each function of the random process the time average is equal to the average for the set of observations.

For the expectation, for example, this condition can be written as $P\{X(t,\ \chi_0)^* \equiv M[X(t,\ \chi_i)]\} = 1$, where the asterisk denotes averaging over time. More rigorous definitions can be found in Wiener (1948).

In accordance with the property of ergodicity of a random process, the nonrandom characteristics defined by one of its implementations with a probability of 1 are equal to the characteristics found throughout the ensemble of realizations. An example of a stationary and ergodic process can be the function $X\ (t,\ \chi) = \cos\ [\omega t + \Psi\ (\chi)]$.

In general, the ergodicity of a random process allows to establish the characteristics of the field using only one of the available implementations. It should be remembered that if the original random process is a sum of two random stationary ergodic processes with normal distribution, then the original process will have the same properties.

2.2.4 The Spectrum of a Random Process

Let us analyze the spectra of a stochastic process for the electromagnetic field modeling. We assume that all of the above restrictions imposed on the model of the field are satisfied. There is given a realization of one model over an infinite period of time and it is required to determine its spectrum. Determination of the spectrum by the Fourier series for a periodic function is valid. However, the periodicity is not a required property of field variations. A non-periodic function can be represented by the Fourier integral:

$$X(t,\ \chi_0) = \int_{-\infty}^{\infty} S_x(\omega, \chi_o) \exp(i\omega t) d\omega \tag{2.5}$$

where $S_x(\omega, \chi_o)$ is the spectrum of a random process.

For the existence of the integral (2.5) it is required to have the absolute integrability of each implementation. To perform this test, it would be necessary to assume that each implementation is damped at infinity. This would be contrary to the condition of stationarity of the process, since the dispersion processes have been changing over time. However, there is an example of undamped functions represented by Fourier integral: $X(t) = \cos\ \omega t^2$. But there is no reason to believe that the implementation of all the fields is like this. Then, it can be argued that the Fourier integral is also not suitable for the construction of the spectrum of a random process.

Let us use a more abstract concept of the integral than in relation (2.5):

$$X(t, \chi_0) = \int_{-\infty}^{\infty} \exp(i\omega t) dZ_F(\omega, \chi_o) \tag{2.6}$$

where Z_F (ω, χ_0) is a function of bounded variation, i.e., the path length of the point on the Z-axis when changing ω from $-\infty$ to ∞ is finite. If Z_F (ω, χ_0) is differentiable in ω, the Stieltjes integral is equal to the Fourier integral. Yet the condition of bounded variation is tough enough: a stationary random process may include implementation presented in the form of (2.6). A way out of this difficult situation was specified by T. Kramer, who proposed to consider the Z_F (ω, χ_0) function as another random process with uncorrelated increments and the convergence of the integral (2.6) to understand the mean-square sense.

Then the Stieltjes integral can be rewritten as

$$X(t, \chi) = \int_{-\infty}^{\infty} \exp(i\omega t) dZ_F(\omega, \chi) \tag{2.7}$$

This integral is often called the representation of Kramer. The value Z_F (ω, χ) is called the spectral measure and can be regarded as an extension of the concept of Fourier coefficients on random processes.

The energy spectrum of a random process follows from the representation of Kramer. Indeed, the energy of the process X (t, χ) in a narrow band of frequencies is the variance of the process in this band, which should be equal to the mathematical expectation of the square of its amplitude. The amplitude of the random process is an integrated value of the spectral measure.

Then the expectation of the squared module of the spectral measures is the energy spectrum of a random process. However, in practice, we use the derivative of the energy spectrum in the frequency domain, this is, the spectral density $S_{xx}(\omega, \chi)$, that can be calculated using the well-known techniques. Determination of the energy spectrum in a narrow band of frequencies $\Delta\omega$ by Kramer may be written as:

$$\int_{\Delta\omega}^{I} S_{xx}(\omega, \chi) \cdot d\omega = \mathrm{M}\left[dZ_x^*(\omega, \chi) \cdot dZ_x(\omega, \chi)\right] \tag{2.8}$$

Hereafter, the asterisk denotes complex conjugation.

The spectral density is a very handy feature of the nonrandom process, perhaps the most important in determining the impedances and its gradients.

It is known that the period range of induction soundings has a set of cycling frequencies, e.g., diurnal period and its harmonics. Therefore, a complete model of the field should consist of part of the field that is modeled by stochastic processes, and the amount of the periodic component. Then the total spectral representation of the field should be written in the form (Volkomirskaya et al. 1979):

$$\mathrm{X} = \int_{-\infty}^{\infty} \exp(i\omega t)dZ_{Fx}(\omega, \chi) + \sum_{k=0}^{n} \exp(i\omega t)\mathrm{X}_k$$

Maybe, such a process describes most of the cases that are of interest for sounding purposes.

2.2.5 *Properties of the Spectral Density*

The spectral density of a random process can be defined by its energy spectrum. Using the mean value theorem, relation (2.8) can be rewritten as $\int'_{\Delta\omega} S'_{xx}(\omega)d\omega = \Delta\omega \mathrm{M}\left[dZ^*_{Fx}(\omega, \chi) \cdot dZ_{Fx}(\omega, \chi)\right]$, where the prime denotes a certain mean value of the spectral density in the frequency range $\Delta\omega$. Then (2.8) can be written as:

$$S'_{xx}(\omega) = (\Delta\omega)^{-1}\mathrm{M}\left[dZ^*_{Fx}(\omega, \chi) \cdot dZ_{Fx}(\omega, \chi)\right]$$

This equation is used for determination of the spectral density. The dimension of the spectral density is equal to the square of the amplitude multiplied by the unit time.

If two components of the field, simulating two random processes, X (t, χ) and Y (t, χ), are studied at the same time, then the mutual spectral density is determined by analogy:

$$S_{xy}(\omega) = (\Delta\omega)^{-1}\mathrm{M}\left[dZ^*_{Fx}(\omega, \chi) \cdot dZ_{Fy}(\omega, \chi)\right]$$

The spectral density $S_{xx}(\omega)$ is a real and positive value, while the mutual spectral density $S_{xy}(\omega)$ is the complex value, and $S^*_{xy}(\omega) = S_{yx}(\omega)$.

Relations between the spectral density and autocorrelation functions are described by the Wiener-Khinchin expressions:

$$S_{xx}(\omega) = \frac{1}{2\pi} \int_{-\infty}^{\infty} R_{xx}(\tau) \cdot \exp(i\omega\tau)d\tau$$

$$S_{xy}(\omega) = \frac{1}{2\pi} \int_{-\infty}^{\infty} R_{xy}(\tau) \cdot \exp(i\omega\tau)d\tau$$

These expressions allow us to calculate the spectral density of the correlation function. Functions are called uncorrelated if $R_{xy}(\tau) \equiv 0$. Therefore, the mutual spectral density of uncorrelated functions is equal to zero. Another important feature of the spectral densities is: if X(t) = a·U(t), then $S_{xx}(\omega) = a^2 \cdot S_{uu}(\omega)$, where a is a real number.

2.3 Impedances as the Transfer Functions

2.3.1 Two-Component Analysis. Coherency

Let us consider the simplest, but important for the understanding, case which is used mainly in magnetovariation sounding methods proposed by Berdichevsky et al. (1969), Schmucker (1970) and Banks (1969). Let two orthogonal components of the magnetic and/or electric fields be observed and modeled by the corresponding random processes B(t,χ) and E(t,χ) with all the properties mentioned earlier. Then, these components may be expressed in terms of spectral measures (2.7) of magnetic and electric fields, respectively. The impedance definition problem reduces to the determination of the transfer function between the spectral measures, as follows from the theory of magnetotelluric soundings with scalar impedance.

In terms of random processes, this linear relation can be written as:

$$dE(\omega,\chi) = \mathrm{Z}(\omega) \cdot dB(\omega,\chi) \tag{2.9}$$

Hereafter, we will omit the index F. It is natural to assume that the transfer function Z(ω) is a non-random characteristic and the structure of the medium is determined only at the point of observation.

Let us express relation (2.9) in terms of the spectral density, the values of which can be evaluated with the experimental data. To do this, multiply both sides of (2.9) by the value of the complex conjugate of $dE(\omega, \chi)$, and then repeat this operation, multiplying both sides of the original equation to $dB^*(\omega, \chi)$. As a result, we get the two new equations:

$$dE(\omega,\chi) \cdot dE^*(\omega,\chi) = \mathrm{Z}(\omega) \cdot dB(\omega,\chi) \cdot dE^*(\omega,\chi)$$
$$dE(\omega,\chi) \cdot dB^*(\omega,\chi) = \mathrm{Z}(\omega) \cdot dB(\omega,\chi) \cdot dB^*(\omega,\chi)$$

Take the expectation of both sides of each equation for χ

$$\begin{aligned} \mathrm{M}[dE \cdot dE^*] &= \mathrm{Z}(\omega)\,\mathrm{M}[dB \cdot dE^*] \\ \mathrm{M}[dE \cdot dB^*] &= \mathrm{Z}(\omega)\,\mathrm{M}[dB \cdot dB^*] \end{aligned} \tag{2.10}$$

Hereinafter the arguments of spectral measures are omitted for brevity. In accordance with the definitions of spectral densities, $S'_{xx}(\omega) = (\Delta\omega)^{-1}$ M[dZ^*_{Fx} (ω, χ) · dZ_{Fx} (ω, χ)] and $S_{xy}(\omega) = (\Delta\omega)^{-1}$ M[dZ^*_{Fx} (ω, χ) · dZ_{Fy} (ω, χ)], after multiplying Eq. (2.10) by $(\Delta\omega)^{-1}$ we obtain the relations between the spectral characteristics of the components of the field:

$$\begin{aligned} S_{EE}(\omega) &= \mathrm{Z}(\omega) \cdot S_{EB}(\omega) \\ S_{BE}(\omega) &= \mathrm{Z}(\omega) \cdot S_{BB}(\omega) \end{aligned} \tag{2.11}$$

Equations (2.11) determine a transfer function in excess process. The impedance can be found from any of those equations. In general, there is no evidence to suggest that the prescribed lower impedances must always be equal and therefore we denote them in different ways:

$$Z'(\omega) = S_{EE}(\omega)/S_{EB}(\omega) \text{ and } \mathrm{Z}''(\omega) = S_{BE}(\omega)/S_{BB}(\omega)$$

The values of the transfer functions obtained by different methods are traditionally compared through the coherence Co(ω). It is the square root of the ratio of these functions $[\mathrm{Z}''(\omega)/\mathrm{Z}'(\omega)]^{1/2}$:

$$\mathrm{Co}^2(\omega) = |S_{EB} \cdot S_{BE}|/(S_{BB} \cdot S_{EE}) = \left|S_{EB} \cdot S^*_{EB}\right|/(S_{BB} \cdot S_{EE}) = |S_{EB}|^2/(S_{BB} \cdot S_{EE})$$

Here the equality $S^*_{xy}(\omega) = S_{yx}(\omega)$ was used and the argument from the spectral densities is omitted. The procedure for dividing the functions is taken so that a mutual spectral density was in the numerator and thus equal to zero in the case of uncorrelated processes. Coherence is the value of nonrandom process indeed, it is varying from zero to one similar to the correlation, but in the frequency domain. If the processes are related linearly, $\mathrm{Co}^2 = 1$.

A striking example of a linear relationship is a connection between periodic processes: on one frequency the squared coherence is always equal to 1! This also holds in geophysics: the meaning of a coherence value of one is that the relationship between the two processes is linear, and that their phase difference is not random in a narrow frequency band of frequencies Δω (Bychkov et al. 1975). Consequently, the coherence value should be considered as a measure of linearity in the relation between the spectral measures of the processes.

The situation is similar if the linear system is considered by means of admittances $B_x = Y_{xy} \cdot E_y$ and $B_y = Y_{yx} \cdot E_x$, where the analogue of random processes is the inverse relationship between measures of spectral components of the field: $dB(\omega, \chi) = Y(\omega) \cdot dE(\omega, \chi)$. Having done all the changes discussed above, we get:

$$\begin{aligned} S_{EB}(\omega) &= \mathrm{Y}'(\omega) \cdot S_{EE}(\omega) \\ S_{BB}(\omega) &= \mathrm{Y}''(\omega) \cdot S_{EE}(\omega) \end{aligned}$$

It is clear that the equations $Z'(\omega) = 1/Y'(\omega)$ and $Z''(\omega) = 1/Y''(\omega)$ will be satisfied always, regardless of noise. However, the coherence can now be rewritten, such as: $Co^2 = Z''(\omega) \cdot Y'(\omega)$.

Since the magnitudes of the transfer function are complex values, then the following expressions (Bendat and Piersol 1986) are used to identify their magnitude and phase separately (on an example of impedences):

$$|Z| = |S_{BE}|/S_{BB} = S_{EE}/|S_{BE}|$$

and

$$\mathrm{Arg}\, Z = arctg(Im\, S_{EB}/Re\, S_{EB}) \tag{2.12}$$

This ends the determination of impedance in two-component linear relationships, and their existence in nature is satisfied ($Co^2 = 1$). However, in practice, the coherence value is never equal to unity. Let us consider the reasons for this phenomenon below.

2.3.2 *Signals with Uncorrelated Noise. Shift Error*

Until now it was assumed that a linear relation introduced by the theory is strictly implemented. But, firstly, such a connection may apply only roughly, and secondly, additional signal sources may be present and lead to uncorrelated signals in the recorded data.

We will continue considering the following situation: the registered field component is a sum of two fields: a field for which the linear relationship $E_x = Z_{xy} \cdot B_y$ and $E_y = -Z_{yx} \cdot B_x$ is valid, and a field for which it is not. Then the magnitude of observed fields, $E°(t, \chi)$ and $B°(t, \chi)$, can be represented in the form of random processes expected theoretically and discussed above, $E(t, \chi)$ and $B(t, \chi)$, and the noise $e(t, \chi)$ and $h(t, \chi)$—two different random processes with the restrictions declared above.

Following the superposition of the fields, the spectral measures of registered signals, $dE^o(\omega, \chi)$ and $dB^o(\omega, \chi)$, can be written as a sum of useful signal measures, $dE(\omega, \chi)$ and $dB(\omega, \chi)$, and the corresponding noise measures referred to as $de(\omega, \chi)$ and $db(\omega, \chi)$. The dependency $dE(\omega, \chi) = Z(\omega) \cdot dB(\omega, \chi)$ between the spectral measures of useful signals through an impedance $Z(\omega)$ can be written as:

$$dE^o - de = Z \cdot [dB^o - db] \tag{2.13}$$

Let us multiply this equality by the complex conjugate term with the measured spectral measures: the first one by $dE^{o*} = dE^* + de^*$, and then $dB^{o*} = dB^* + db^*$. As a result, we get two new equations:

$$dE^{o} \cdot dE^{o*} - de \cdot (dE^{*} + de^{*}) = Z \cdot [dB^{o} \cdot dE^{o} - db \cdot (dE^{*} + de^{*})];$$
$$dE^{o} \cdot dB^{o*} - de \cdot (dB^{*} + db^{*}) = Z \cdot [dB^{o} \cdot dB^{o*} - db \cdot (dB^{*} + db^{*})].$$

To move from these equations to the spectral densities, it is necessary to take the operation of the expectation of both sides of these equalities. We thus assume that the noise is not correlated either with each other or with the useful signal, i.e., the equalities

$$\mathrm{M}[dE^{*} \cdot de] = \mathrm{M}[dE^{*} \cdot db] = \mathrm{M}[de^{*} \cdot db] = \mathrm{M}[dH^{*} \cdot de] = \mathrm{M}[dH^{*} \cdot db] = 0$$

In such a situation, the part of input signal will not be correlated with the part of output signal. So, taking the mathematical expectation of the operation of the two original equations and taking into account the equalities written above, we go to the spectral density:

$$\begin{aligned} S^{o}_{EE} - S_{ee} &= Z(\omega) \cdot S^{o}_{EB} \\ S^{o}_{BE} &= Z(\omega) \cdot \left(S^{o}_{BB} - S_{bb}\right) \end{aligned} \tag{2.14}$$

The impedance Z(ω) cannot be unequivocally defined from these equations because the noise intensities, S_{ee} and S_{bb}, are not known. In this situation, the measured values are used to determine the spectral densities of the impedance via the following relations, as if there were no noise at all:

$$Z'(\omega) = S^{o}_{EE}/S^{o}_{EB} \quad \text{and} \quad Z''(\omega) = S^{o}_{BE}/S^{o}_{BB} \tag{2.15}$$

Of course, different impedance estimations are obtained, and this deviation is called shift error. It is important to remember that, in practice, this failure may significantly exceed the random error and does not decrease with increasing number of raw data. Then, the estimate of the coherence of the same measured spectral densities should be calculated:

$$\mathrm{Co}^{2} = |S_{BE}|^{2}/(S_{EE} + S_{ee}) \cdot (S_{BB} + S_{bb}).$$

Here, the ratio $S_{xy}* = S_{yx}$ was used. It can be seen than Co^2 is equal to one only when $S_{ee} = S_{bb} = 0$. If $\mathrm{Co}^2 < 1$, the impedances have a different assessment.

It is useful to determine how a true value of the impedance Z(ω) can be connected with its measure Z′(ω) and Z″(ω). Taking into account that $S^{o}_{BE} = S^{o}_{EE}/Z' = Z'' \cdot S^{o}_{BB}$, we will rewrite Eq. (2.14) in the form:

$$\begin{aligned} Z'(\omega) &= Z(\omega)/\left[1 - S_{ee}/S^{o}_{EE}\right] \\ Z''(\omega) &= Z(\omega) \cdot \left[1 - S_{bb}/S^{o}_{BB}\right] \end{aligned}$$

Since the spectral density of noise is obviously less than the sums of the measured signals, the assessment of the impedance $Z'(\omega)$ is overstated, and that of $Z''(\omega)$—understated in comparison with the true impedance value, so the following inequality is valid:

$$|Z'(\omega)| > |Z(\omega)| > |Z''(\omega)|$$

It should be emphasized that the presence of uncorrelated noise leads to differences in the modules of estimates only. Estimates of the argument are equal to the true impedance argument, as constructed by unbiased assessment of the mutual spectral density S_{BE} (Lejbo 1978). This fact can offer a method of determining the impedance using only unshifted estimates of cross-spectral densities (remote reference method). It is enough to multiply the original Eq. (2.13) by the spectral measures of signal from a remote observation point with uncorrelated noise *de*' and *db*'. This gives the zero value of S_{ee} and S_{bb}.

It is possible to reduce the shift of the impedance module by considering, for example, the geometric mean of the two discussed above $Z^{\times} = [|Z'| \cdot |Z''|]^{1/2}$. This estimate coincides with the true impedance of the module if the ratios of "signal/noise" are the same on both channels. Communication of such estimates can be obtained through the coherence also as the ratio of first two impedances from the relation above. Then $|Z^{\times}| = |Z'| \cdot \text{Co}$ and $Z^{\times}| = |Z''|/\text{Co}$, that is often used in practice. In general, estimate $Z^{\times}$ has a smaller displacement than the original, but its value is unknown.

All the relations presented above can be applied to the admittances. Moreover, based on the type of impedance and admittance estimates, it can be argued that the estimate $Z^{\times}$ is determined by other formulas too: $Z^{\times} = [|Z''| \cdot |Y'|]^{1/2} = [|Z'| \cdot |Y''|]^{1/2}$. Thus, the estimate $Z^{\times}$ is the geometric mean of the corresponding impedance and admittance estimates. All these formulas are valid when the two processes are analyzed with uncorrelated noise.

The estimates of admittances and impedances, as described above, depend on the noise level for each of the two channels that are controlled by the coherence. Then the interpretation of any assessment can be carried out only if its value is high enough. The first two noise estimates depend on the intensity of only one of the channels. They can be used with a priori information about the noise power on each channel. For example, in the range of long periods the noise is usually concentrated in the electric field and a reliable estimate will be S^{o}_{EB}/S^{o}_{BB}. Phases of impedances and admittances are the most reliable data for interpretation.

More generally, the observed electromagnetic field is a sum of fields used for soundings and other fields which can also be linked together, for example, linearly, i.e. $de = \eta(\omega) \cdot db$. The sources of the fields are different and we will consider them to remain uncorrelated. Then the values of cross-spectral density of noise, S_{eb} and S_{be}, are not zero, and after the capture of the expectations of the spectral measures, instead the equality (2.14) we will obtain the equations in the form:

$$S^{o}_{EE} - S_{ee} = Z(\omega) \cdot \left(S^{o}_{BE} - S_{be}\right)$$
$$S^{o}_{EB} - S_{eb} = Z(\omega) \cdot \left(S^{o}_{BB} - S_{bb}\right)$$

Since the intensity of the noise is unknown beforehand, we shall seek the transfer functions, as before, in the form of relationships, not taking into account the noise (2.15). According to $S_{be} = S_{ee}/\eta(\omega)$ and $S_{eb} = \eta(\omega)S_{bb}$, we get:

$$Z'(\omega) = Z(\omega)/\left[1 - S_{be}/S^{o}_{HE}\right] + S_{be} \cdot \eta(\omega)/S^{o}_{BE}$$
$$Z''(\omega) = Z(\omega)\left[1 - S_{bb}/S^{o}_{BB}\right] + \eta(\omega) \cdot S_{bb}/S^{o}_{BB}$$

The last equation shows that if $\eta(\omega) = Z(\omega)$, i.e. both sources are the same, then the result of sounding will not differ from the case of a single source without noise. If these two transfer functions are different, then not only the module but also the phase of the impedance value will be shifted.

So, a brief analysis of the impedance estimates in the presence of correlated noise with unknown transfer function indicates a lack of baseline information for any practical conclusions on the shifted values of estimates. So, it gives the opportunity to make the soundings. However, this approach makes it possible to perform the soundings in the presence of several field sources.

2.4 Data Processing of Dst Field Components

The theoretical principles of magnetovariation sounding dealt with electromagnetic fields induced in the ionosphere and/or magnetosphere of the Earth, i.e., with the external sources which are used for these electromagnetic soundings. However, in practice a lot of electromagnetic fields of other kind can be registered simultaneously (Junge 1986). In this situation, we have to treat the registered field components as a sum of the signals and noise. The main purpose of the data processing is to separate these sources and to estimate the transfer functions corresponding to the responses considered in the theory.

Let us analyze the simplest situation with two measured field components, as it usually done in the magnetovariation method. Assuming H and Z to be the signal spectra (Fourier transform) of the horizontal and vertical magnetic field components, and h and z, the noise spectra of the corresponding measured components, we can write the observed spectra in the following form:

$$\widetilde{H} = H + h \quad \text{and} \quad \widetilde{Z} = Z + z$$

Then, the searched response function $W(\omega) = B_Z/B_H$ in equation $Z = i\omega\mu/2 \cdot R \cdot tg\theta \cdot B_Z/B_H$ can be written in our case as:

$$(\widetilde{Z} - z) = W \cdot (\widetilde{H} - h) \tag{2.16}$$

Let us multiply both parts of Eq. (2.16) by complex conjugate values of the measured field components: first by $\widetilde{Z}^* = Z^* + z^*$ and then by $\widetilde{H}^* = H^* + h^*$. In consequence, we will obtain two equations which can be written in the form:

$$\begin{aligned} \widetilde{Z} \cdot \widetilde{Z}^* - (Z^* + z^*) \cdot z &= W \cdot \left[\widetilde{H} \cdot \widetilde{Z}^* - (Z^* + z^*) \cdot h\right], \\ \widetilde{Z} \cdot \widetilde{H}^* - (H^* + h^*) \cdot z &= W \cdot \left[\widetilde{H} \cdot \widetilde{H}^* - (H^* + h^*) \cdot h\right]. \end{aligned} \tag{2.17}$$

To obtain the spectral densities it is necessary to average the values like $\widetilde{H} \cdot \widetilde{H}^*$ through several time series or to smooth them through several frequencies. Both procedures are equal for linear analysis and yield the spectral density estimation S, which can be written in the form

$$S_{ij}(\omega) = (\Delta\omega)^{-1} \left\langle H_i(\omega, \kappa) \cdot H_j^*(\omega, \kappa) \right\rangle \tag{2.18}$$

Here κ is the realization number in a simplest case, $\langle \ldots \rangle$ is the averaging over κ or smoothing operation over ω. More details were considered by Semenov (1985), with the explanation why this method cannot be used for regular oscillation like Sq. If H_i and H_j are not correlated, the spectral density will be equal to zero. Using Eq. (2.18), expression (2.17) can be rewritten for spectral densities as follows:

$$\widetilde{S}_{ZZ} - S_{zz} = W \cdot \widetilde{S}_{HZ} \quad \text{and} \quad \widetilde{S}_{ZH} = W \cdot (\widetilde{S}_{HH} - S_{hh}) \tag{2.19}$$

It was taken into consideration here that both noises are not correlated with signals and with each other, i.e., we had:

$$\langle Z^* \cdot z \rangle = \langle Z^* \cdot h \rangle = \langle z^* \cdot h \rangle = \langle H^* \cdot z \rangle = \langle H^* \cdot h \rangle = 0$$

It is obvious that transfer function W cannot be found from Eq. (2.19) because the noise densities S_{zz} and S_{hh} are unknown. In this situation we can only estimate the transfer function. For this purpose, let us assume that the noises are neglected. Then, from two Eq. (2.19) without noise spectra we can determine two estimates:

$$W_1 = \widetilde{S}_{ZZ} / \widetilde{S}_{HZ} \quad \text{and} \quad W_2 = \widetilde{S}_{ZH} / \widetilde{S}_{HH} \tag{2.20}$$

To verify the justification of our assumption, let us consider the ratio of these two estimates of W:

$$\frac{W_2}{W_1} = \frac{\widetilde{S}_{ZH} \cdot \widetilde{S}_{HZ}}{\widetilde{S}_{HH} \cdot \widetilde{S}_{ZZ}} = \frac{\left|\widetilde{S}_{ZH}\right|^2}{\widetilde{S}_{HH} \cdot \widetilde{S}_{ZZ}} = \mathrm{Co}^2 \tag{2.21}$$

Quantity Co^2 is named a common coherence function which is a measure of the linear relationship in the frequency domain. Using expressions with noises it is easy to show that $\mathrm{Co} \leq 1$. If Co = 0, relation (2.16) cannot describe the measured field, i.e. only uncorrelated noises were recorderd. If Co = 1, both estimations, W_1 and W_2, are equal and noises are absent.

In fact, $1 \geq \mathrm{Co} \geq 0$, and it is interesting to establish how the searched response function is related to its estimation, W_1 and W_2. From Eqs. (2.19) and (2.20) it is easy to show that $W_2 = W_1 \cdot (1 - \widetilde{S}_{hh}/\widetilde{S}_{HH})$ and $W_1 = W/(1 - \widetilde{S}_{zz}/\widetilde{S}_{ZZ})$. Assuming the noises less than the signals: $\left|S_{zz}/\widetilde{S}_{ZZ}\right| < 1$ and $\left|S_{hh}/\widetilde{S}_{HH}\right| < 1$, we have:

$$|W_1(\omega)| \geq |W(\omega)| \geq |W_2(\omega)| \qquad (2.22)$$

i.e., the searched value lies between the two estimates. However, this fact is valid only for module values. The phases of both estimations are equal:

$$\mathrm{Arg}\, W_1 = \mathrm{arc\,tg}\,(Im\widetilde{S}_{ZH}/Re\widetilde{S}_{ZH}) = \mathrm{Arg}\, W_2 \qquad (2.23)$$

This is the reason why we can say that the phase estimations are more reliable than the module ones for the magnetovariation sounding. These phase estimations are free from the shift errors, a characteristic of which is the coherence.

2.5 Principal and Selected Directions in Magnetotelluric

Many methods were suggested to choose the principal directions in the magnetotelluric investigation (Yee and Paulson 1987). As a rule, these methods were based on the selection of directions in which the additional impedance has a minimum. However, the spectral analysis cannot estimate zero. From the point of view of this kind of data processing, two possibilities exist:

$$E_y = Z_m \cdot B_x + Z_a \cdot B_y, \quad \text{or} \quad E_y = Z_m \cdot B_x + e. \qquad (2.24)$$

Here E_y is the electric field component in the direction of the B_y magnetic component, B_x is the magnetic field component perpendicular to both E_y and B_y components, Z_m and Z_a are main and additional impedances, respectively.

The difference between these equivalencies is obvious: the term $Z_a \cdot B_y$ is replaced by the noises e in the second equivalence. Then, instead of attempts to find $Z_a = 0$, we can establish which one of the relationships of Eq. (2.24) we are dealing with. It was shown (Semenov and Kaikkonen 1986) that the second equivalence can be considered instead of the first one provided that there is the equality to zero of common coherency, $\mathrm{Co}_{BB} = 0$, and partial coherency, $\mathrm{Co}_{EB/B} = 0$, simultaneously. This means that under these conditions the measured component E_y correlates only with the component B_x which is not related with component B_y. This corresponds to the zero value of the additional impedance in the deterministic theory. If those coherence functions are zero for one or two nonorthogonal directions, we have found the selected directions. If both selected directions are orthogonal, we have found the principal directions.

As it was mentioned above, almost all kinds of coherence functions are not invariant during the reference system rotation (Bath 1974). This fact allows us to use the diagrams of these coherencies.

2.6 Confidence Limits

Many methods exist to estimate the spectral densities (2.18), which differ as a rule in the calculations algorithms. However, the general characteristic exists for all these methods: it is the degree of freedom ν. For example, for smoothing through several frequencies, ν is twice the number of them, for the averaging of the Fourier transforms for several time series, ν is twice the number of the time series used. The same value can be calculated from other methods. Some special case is the analysis in the time domain only (Wieladek and Ernst 1977; Svetov and Shimilevich 1982).

Let any estimation of the response Z be found in the narrow frequency range known from the analysis. Then the confidence limits can be estimated according to the formulae (Bendat and Piersol 1986):

$$\begin{aligned} \delta\left|\widetilde{Z}\right| &= \frac{2q}{\nu - 2q} \cdot \mathbf{F} \cdot \frac{\left[1 - Co_{EBB}^2\right] \cdot \widetilde{S}_{EE}}{\left[1 - Co_{BB}^2\right] \cdot \widetilde{S}_{BB}} \\ \delta\,\mathrm{Arg}\,\widetilde{Z} &= arc\,\sin\left(\delta\left|\widetilde{Z}\right| / \left|\widetilde{Z}\right|\right), \end{aligned} \tag{2.25}$$

where q is the number of input signals (two in our case), **F** is Fisher's parameter, $\mathrm{Co_{BB}}$ is the common coherency between inputs and $\mathrm{Co_{EBB}}$ is the multiple coherency.

Theoretically, expression (2.25) is valid for the case of normal distributions of the field intensities. Formula (2.25) estimates the random errors, while the coherencies estimate the possible shift error.

2.7 Mean and Robust Estimations

The task to average the obtained results for impedances, admittances and apparent resistivities appears very often in many situations. The apparent resistivity is connected with the impedance by formula presented in Chap. 1.

Let us recall that all transfer functions are complex ones. Then the mean complex values can be written for two apparent resistivities ρ (for one direction, for example) in the form:

$$\bar{\rho} = \sqrt{(\rho_1 \cdot \rho_2)} = \sqrt{(|\rho_1| \cdot |\rho_2|)} \cdot e^{-i(\varphi_1 + \varphi_2)/2}. \tag{2.26}$$

As long as we are going to average the modules and phases separately, the mean geometrical values of the modules will be in accordance with the arithmetical mean values of phases.

As a response function, the real and imaginary parts of C are widely used too. They are connected with the apparent resistivity by the formulae:

$$\begin{aligned} \rho &= \left[(ImC)^2 + (ReC)^2\right] \cdot 0.8 \cdot \pi^2 \cdot f, \\ Arg\,\rho &= \pm 2 \cdot \text{arc tg}\left(\frac{ImC}{ReC}\right) + 90^\circ \end{aligned} \tag{2.27}$$

Here ρ is expressed in Ohm-m, C in km, f is frequency in cycles per second, Arg ρ is in degrees and the sign in the second expression depends on the assumed oscillation term exp($\pm i\omega t$). For example, this sign is minus for data of Schultz and Larsen (1987) and it is plus for data of Roberts (1984). The mean values of the C responses can be considered as the arithmetic mean values: $\overline{ReC} = (ReC_1 + ReC_2)/2$, for example.

The other way to determine the averaged value of response functions is robust estimation. This technique uses the absolute values as the error criteria, instead of the least-squares ones (Claerbout and Muir 1973). The idea is simple: a set of N real samples $|\rho_i|$ or φ_i may be sorted in the order:

$$\rho_1 \le \rho_2 \le \cdots \le \rho_m \le \cdots \rho_{N-1} \le \rho_N \tag{2.28}$$

The median (middle value ρ_m) can be different from the arithmetical or geometric means due to several blunders. In this case, to construct the error bars, the following differences are considered:

$$\Delta\rho_i = |\rho_i - \rho_m|,$$

which are also sorted in the order $\Delta\rho_1 \le \Delta\rho_2 \le \cdots \Delta\rho_m \le \cdots \Delta\rho_{N-1} \le \Delta\rho_N$.

The middle values of this set is named the median absolute deviation (MAD). Then the estimation of ρ can be written as follows (with the theoretically known probability)

$$\rho_m + \text{MAD} \ge \rho \ge \rho_m - \text{MAD}$$

The question arises how to improve this set of ρ_i in order to reduce the MAD value. It is obvious that the highest errors are concentrated at edges of the row

(2.28). Then, if several samples are removed from both edges, we can reach the aim. We have a possibility to make this process iteratively. As a criterion of finishing these iterations, the approximate coincidence of the mean and median estimations for a new row can be considered. More details can be found in Egbert and Booker (1986) and Chave et al. (1987).

The apparent resistivities or impedances averaged by different methods may not coincide exactly with each other. This can be the reason why the same initial responses can produce somewhat different geoelectrical structures even if the conditions of inversion are the same.

2.8 Conclusions

In this chapter we have considered the principles how to estimate solutions of one equation with several unknown values by means of statistical methods.

References

Banks, R.J.: Geomagnetic variations and the electrical conductivity of the upper mantle. Geophys. J. Roy. Astron. Soc. **17**, 457–487 (1969)

Bath, M.: Spectral Analysis in Geophysics. Elsevier, Amsterdam (1974)

Bendat, J., Piersol, A.: Random Data Analysis and Measurement Procedures, 2nd edn. Wiley, New York (1986)

Berdichevsky, M.N., Vanyan, L.L., Feinberg, E.B.: The frequency sounding of the Earth using spherical analysis results of geomagnetic variations. Geomag. Aeron. **9**, 372–374 (1969) (in Russian)

Bychkov, V.S., Lejbo, A.B., Semenov, V.: Interpretation of coherency coefficient and phase difference function. Geomagnetic Studies, 16. Moscow, Nauka (1975) (in Russian)

Chave, A.D., Thomson, D.J., Ander, M.E.: The robust estimation of power spectra, coherencies, and transfer functions. J. Geophys. Res. **92**(B1), 633–648 (1987)

Claerbout, J.F., Muir, F.: Robust modelling with erratic data. Geophysics **38**(5), 826–844 (1973)

Egbert, G.D., Booker, J.R.: Robust estimation of geomagnetic transfer functions. Geophys. J. Roy. Astron. Soc. **87**, 173–194 (1986)

Jenkins, G.M., Watts, D.G.: Spectral Analysis and Its Application. Holden-Day, San-Francisco (1968)

Junge, A.: Characterization of and correction for cultural noise. Surv. Geophys. **17**, 361–391 (1986)

Lejbo, A.B.: Probabilistic model for estimating the impedance in a horizontally-homogeneous media. Geomag. Aeron. **20**(4), 706–710 (1978) (in Russian)

Reddy, I.K., Rankin, D.: Coherence functions for magnetotelluric analysis. Geophysics **39**(3), 312–320 (1974)

Roberts, R.G.: The long period electromagnetic response of the Earth. Geophys. J. Roy. Astron. Soc. **78**(2), 547–572 (1984)

Semenov, V.: Data processing of magnetotelluric soundings/Obrabotka dannyh magnitotelluricheskogo zondirovaniya. Moscow, Nedra, 133 p. (1985) (in Russian)

Semenov, V., Kaikkonen, P.: On magnetotelluric data analysis with example for the Kuhmo region in Eastern Finland. Report 13 of the Oulu University, Finland, 1–13 (1986)
Svetov, B.S., Shimilevich, 1982. Determination of the linear relation between the components of magnetotelluric field. Izv. AN USSR, Fizika Zemli, 5, 59–67 (1982) (in Russian)
Schmucker, U.: Anomalies of geomagnetic variations in the southwestern United States. Bull. Scripps Inst. Ocean. Univ. Calif. 1–165
Schultz, A., Larsen, J.C.: On the electrical conductivity of the mid-mantle: I. Calculation of equivalent scalar magnetotelluric response function. Geophys. J. Roy. Astron. Soc. **88**, 733–761 (1987)
Volkomirskaya, L.B., Lejbo, A.B., Semenov, V.: On the statistical treatment of the electromagnetic sounding data. Geomag. Aeron. **19**(5), 530–537 (1979) (in Russian)
Wiener, N.: Cybernetics: Or Control and Communication in the Animal and the Machine. The M.I. T Press, Cambridge, Massachusetts (1948)
Wieladek, R., Ernst, T.: Application of the method of least squares to determining impulse responses and transfer functions. Publ. Inst. Geophys. Pol. Acad. Sci. G-1, **110**, 3–12 (1977)
Yee, E., Paulson, K.V.: The canonical decomposition and its relationship in other forms of magnetotelluric impedance tensor analysis. J. Geophys. **63**, 173–189 (1987)

Chapter 3
Modeling of Deep Soundings

Abstract The impedances for the deep electromagnetic soundings of the Earth are obtained from the relations for the Fourier amplitudes of the observed field components. These relations are essentially different for the magnetotelluric and magnetovariation sounding methods. In order to increase the reliability of investigations, studies of electrical properties of the Earth's mantle are often carried out by the joint inversion of impedances obtained by both methods of sounding. The forward modeling is a unique tool to verify the accuracy of merging the differently obtained long-period impedances because those simplified relations were derived theoretically for the radio-wave periods only. The spherical modeling of the responses above 2D mantle inhomogeneities presented in this paper has shown that the induction methods can give mutually inconsistent results and the combinations of their responses can be problematic in practice. For this reason, much attention is given to the generalized magnetovariation sounding method which results in both, regularized impedance functions in space and frequency domains and closely resemble the magnetotelluric ones devoid of the subsurface galvanic distortions. In this study, some peculiar properties of the induction arrows above a spherical inhomogeneity excited by an inhomogeneous external field are estimated for long periods. The final comprehensive model, assuming a shell of the realistic Earth's surface conductance, is an evidence that the generalized magnetovariation method is promising for the study of mantle inhomogeneities and can be used with the magnetotelluric method in a specific way.

Keywords Numerical simulation · Mantle step

The chapter is an amended version of the paper: Vozar, J., and V. Y. Semenov (2010), Compatibility of induction methods for mantle soundings, J. Geophys. Res., 115, B03101, doi:10.1029/2009JB006390.

© Springer International Publishing AG 2018

V. Semenov and M. Petrishchev, *Induction Soundings of the Earth's Mantle*,
GeoPlanet: Earth and Planetary Sciences, DOI 10.1007/978-3-319-53795-5_3

3.1 Introduction

The induction soundings of the Earth's mantle are based on interpretation of the response functions obtained by both, the magnetotelluric sounding, referred as MTS, and the generalized magnetovariation sounding, denoted as MVS. The theoretical details have been considered in Sect. 1.5.

The presented chapter attempts to provide an image how the generalized MVS method works on synthetic data prior to its wide application on real data in order to test the reliability of the proposed approach. Another aim of the analysis is to point out the advantages of the MVS method in comparison with the methods currently used. We believe that this study, initiated by the already cited works by U. Schmucker and V. Shuman, could help to reach higher confidence in the mantle sounding results and to achieve as broad a depth range of the mantle as possible.

3.2 Numerical Simulations of Induction Soundings

Forward modeling of electromagnetic fields excited by the ionospheric and magnetospheric sources has been carried out on the globe with geographic and geomagnetic reference systems aligned. The calculated field components were used for determining the impedances for the MTS and MVS methods, in accordance with the corresponding IBC's.

The assumed layered Earth's structure with a sudden decrease (depth step) of the highly conductive top layer of the upper mantle (asthenosphere) is shown in Fig. 3.1. This sharp (but not discontinuous) step in depth will provide a possibility of estimating the maximal effects on the fields induced by this mantle inhomogeneity. Spatial distributions of all field components have been obtained on the globe at a grid of $1° \times 1°$ for a period range from 10 min up to 4096 days. Basically, three spherical models were used for testing the method: 2D and 3D spherical models with the homogeneous surface conductance of 20 S assumed, and 3D spherical model where the realistic surface shell conductance has been taken into consideration.

There are no magnetotelluric source plane waves for spherical models. But, since we work only with impedances, it is sufficient to have horizontal (tangential) fields that locally depend linearly on the horizontal coordinates (Berdichevsky and Dmitriev 2008). So, one can excite the modeled Earth by three polarizations and thus obtain a tensor of impedances. One needs three polarizations (instead of two in the plane case) in order to avoid the singularities arising globally due to the change of signs of *cos* and *sin* functions. A spherical analog for the "plane wave" source used in the MTS method can be approximated by three orthogonal ionospheric sources of ring current type. A single magnetospheric ring current is sufficient as a source for modeling in the MVS methods based on the corresponding relations mentioned above (Kuvshinov et al. 2005).

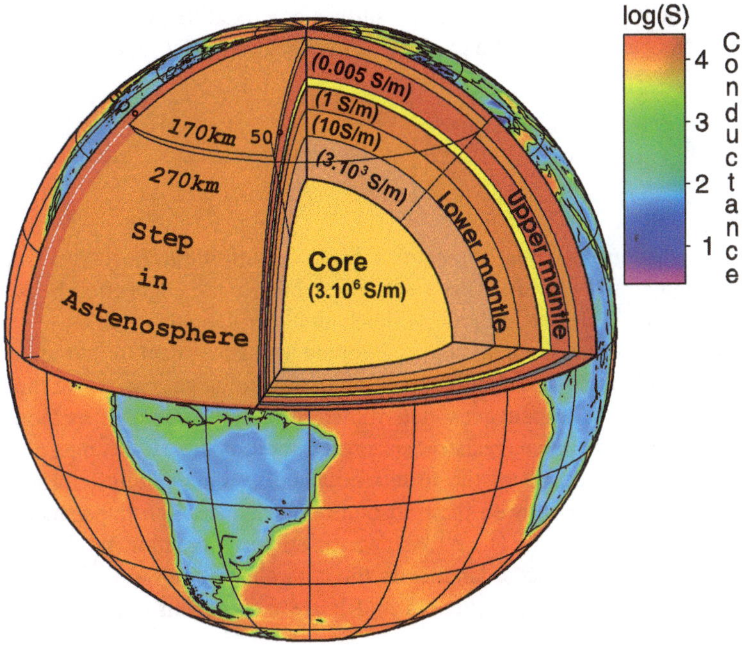

Fig. 3.1 The schematic model of the Earth's interiors assumed in our modeling with the surface shell conductance

Following Guglielmi (1989), the MVS relation (1.16) has been considered as a differential equation with unknown scalar impedance:

$$\begin{aligned}(B_\theta/R)\cdot\partial\zeta_m/\partial\theta + [B_\phi/(R\sin\theta)]\cdot\partial\zeta_m/\partial\phi \\ +\zeta_m[\partial(B_\theta\cdot\sin\theta)/\partial\theta + \partial B_\phi/\partial\phi]/(R\cdot\sin\theta) - i\omega B_r = 0\end{aligned} \tag{3.1}$$

The general solution of Eq. (3.1) on the surface has been obtained for the spherical 3D inhomogeneity by the numerical finite difference method: a simple five-point stencil discretization was applied to write central finite difference approximations of derivatives at spherical grid points. As a result, a system of linear equations (with a small modification of the stencil on the grid boundary to an asymmetric one) has been obtained with the impedances as unknowns at all grid points.

The general Eq. (3.1) for the 3D case was simplified for the 2D axially symmetric conductivity distribution:

$$(B_\theta/R)\cdot\partial\zeta_m/\partial\theta + [\partial(B_\theta\cdot\sin\theta)/\partial\theta]/(R\cdot\sin\theta)\cdot\zeta_m - i\omega B_r = 0 \tag{3.2}$$

Its analytical solution on the Earth's surface at a fixed frequency independent of longitude can be written in the explicit form (Semenov et al. 2007):

$$\zeta_m(\theta,\varphi_0,\omega)=e^{-\int\limits_{\theta_0}^{\theta}a(\theta,\varphi_0,\omega)d\theta}\left\{\int\limits_{\theta_0}^{\theta}b(\theta,\varphi_0,\omega)\,e^{\int\limits_{\theta_0}^{\theta}a(\theta,\varphi_0,\omega)d\theta}d\theta+C_{m0}\right\}$$

where $C_{\mathrm{m0}} = \zeta_{\mathrm{m}}(\theta_0, \varphi_0, \omega)$, $a(\theta, \varphi, \omega) = [\partial(H_\theta \cdot \sin\theta)/\partial\theta]/(H_\theta \cdot \sin\theta)$, and $b(\theta, \varphi_0, \omega) = i\omega\mu_0 \cdot R \cdot H_r/H_\theta$. This solution establishes the connection between the impedance values at two different points along co-latitude profiles. A starting (reference) point was chosen at the quasi homogeneous segment to find impedances and consequently their spatial derivatives along the profile passing above the mantle inhomogeneity. The impedances determined from all sounding methods were converted into the apparent resistivities in the traditional way, keeping the impedance phases. Comparison of the results obtained by the analytical formula (3.2) with the formula (3.1) showed their absolute coincidence. The impedance surface distribution found from (3.1) or (3.2) was used for estimation of the induction arrows on the sphere in accordance with the common definition: $C_u = (\mathrm{Re}\{A\}; \mathrm{Re}\{B\})$, $C_v = (\mathrm{Im}\{A\}; \mathrm{Im}\{B\})$, where A and B are the gradient tippers:

$$A=[1/(i\omega\,\mu_0 R)]\cdot\partial\zeta_m/\partial\theta;\,B=[1/(i\omega\mu_0\,R\sin\theta)]\cdot\partial\zeta_m/\partial\phi$$

In the simplified case of axially symmetric anomaly, the gradient tippers have only one component, $C_u = (\mathrm{Re}\{A\}; 0)$, $C_v = (\mathrm{Im}\{A\}; 0)$:

$$C_u=(\mathrm{Re}\{B_r-\{[\partial(B_\theta\cdot\sin\theta)/\partial\theta]/(R\cdot\sin\theta\cdot i\omega\,\mu_0)\}\cdot\zeta_m\}/B_\theta);0),$$
$$C_v=(\mathrm{Im}\{B_r-\{[\partial(B_\theta\cdot\sin\theta)/\partial\theta]/(R\cdot\sin\theta\cdot i\omega\,\mu_0)\}\cdot\zeta_m\}/B_q\};0).$$

3.3 Modeling Results

The first one of the mentioned models, the axially symmetric model of the Earth, is shown in Fig. 3.1. The mantle layer (asthenosphere) of 100 km thickness with a conductance of 10 kS was situated 170 km below the surface at co-latitudes of less than 40° measured from the North Pole. In the residual part of the model, it was considered to be at the 270 km depth. The source excitations were assumed as specified above. The MTS and MVS responses on the surface just above the above-mentioned layer decrease were compared with the horizontal spatial gradients (HSG, see (1.6) in Sect. 1.2) and geomagnetic depth sounding (GDS, see (1.7) in Sect. 1.2) ones. Figure 3.2 displays the apparent resistivities and impedance phases in the period range from 360 s to 3 months at a chosen point just above the layer decrease (left side). The right side of Fig. 3.2 depicts the induction vector module, apparent resistivity and impedance phase changes along the profile from

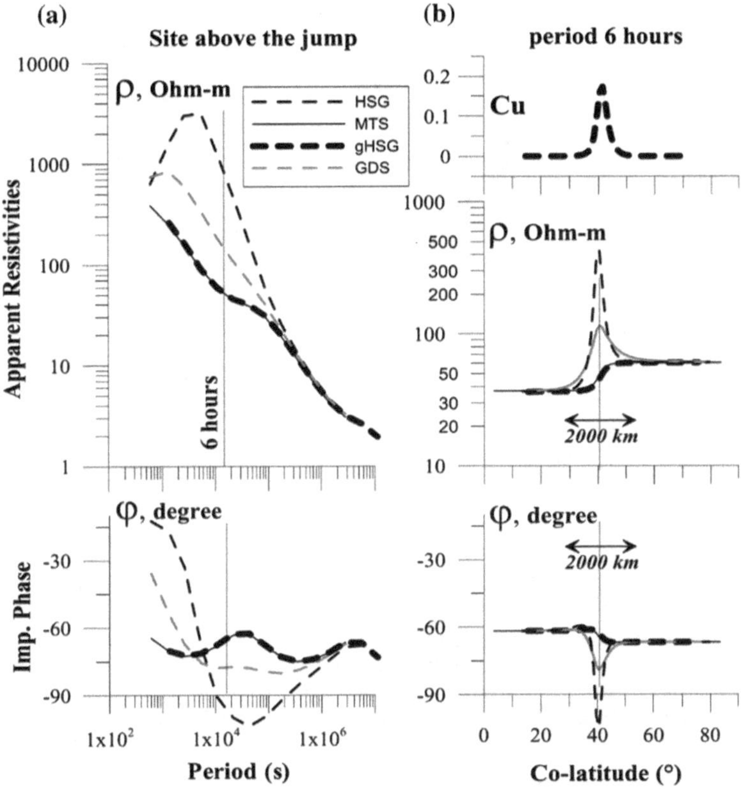

Fig. 3.2 Apparent resistivities and impedance phases just above the layer depth stepin the upper mantle as a function of periods (*left side*) and the profile from the North Pole to the equator (*right side*), with real induction arrows for a period of 6 h. The responses of the E-polarized MTS (*thin solid*) are also shown. The HSG, GDS and MVS methods are designated as thick and thin dashed curves, respectively

the North Pole to the equator for the period of six hours. It is clear that the MTS and MVS responses are practically the same along the whole profile and for all the periods. However, the HSG and GDS responses are completely dissimilar at shorter periods (up to approximately 10^5 s) and along a distance of about 2000 km where the induction arrows appear; see the right side of Fig. 3.2.

An analogous model was prepared for a deeper conductive mid-mantle layer with a thickness of 150 km and conductance of 300 kS. The upper surface of this layer drops from 500 to 650 km depth at the same spherical coordinates θ and φ as in the first model, as well as the same source excitations have been assumed. The MTS and MVS responses are again practically the same for the period range from 0.25 days up to 11 years and along the same profile as for the first model; see Fig. 3.3. The GDS responses differ from them for periods up to 10^7 s but the curve continuances are similar, which is a consequence of the inhomogeneity's greater

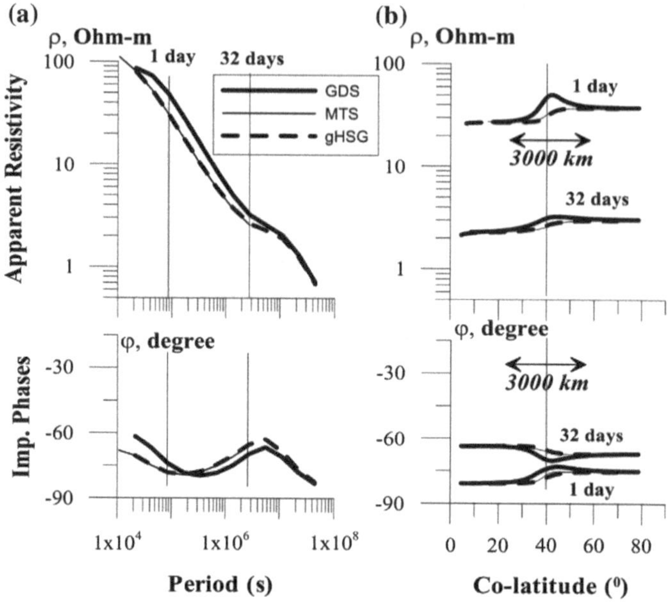

Fig. 3.3 **a** Apparent resistivities and impedance phases, directly above the stepin the upper mantle, as a function of period. **b** A colatitude profile from the North Pole to the equator, for the period 1 day and 32 days. The responses of the E-polarized MT sounding (MTS) and the GDS and generalized HSG (gHSG, see Sect. 1.5) methods are shown, as indicated in the legend

depth as compared to the first model; see the left side of Fig. 3.3. Apparent resistivities and impedance phases obtained for MTS and MVS differ from the ones for GDS along a distance of about 3000 km above so deeply located mantle inhomogeneity; see the right side of Fig. 3.3.

The induction arrows on the Earth's surface presented in Fig. 3.4 are shown just above the sudden depth stepof the upper mantle layer described by the first model. They are induced by the axially symmetric conductive anomalies situated: (a) along the co-latitude 40° (latitude 50°), (b) along the co-latitude 90° (latitude 0°), and (c) along longitude 0°. They were computed numerically by the finite difference method applied to the differential relation (3.1).

The results indicate quantitative changes of the induction arrows computed on the Earth's surface along longitudes and latitudes, depending on the position of inhomogeneity. Arrows disappear near the poles and achieve maximal values near the equator, which is an unusual effect as compared with the common arrows considered in the frame of the plane model. The obtained effect is connected with the sphericity of the model. Here, the impedance gradients are calculated for the degree mesh instead of the one of uniform distance for the case of the flat Earth model. Note that the model has been calculated only with inhomogeneities of finite dimensions in order to avoid possible singularities at the poles and their vicinity. As a result of sphericity and the assumed model symmetry, dimensions of the anomaly

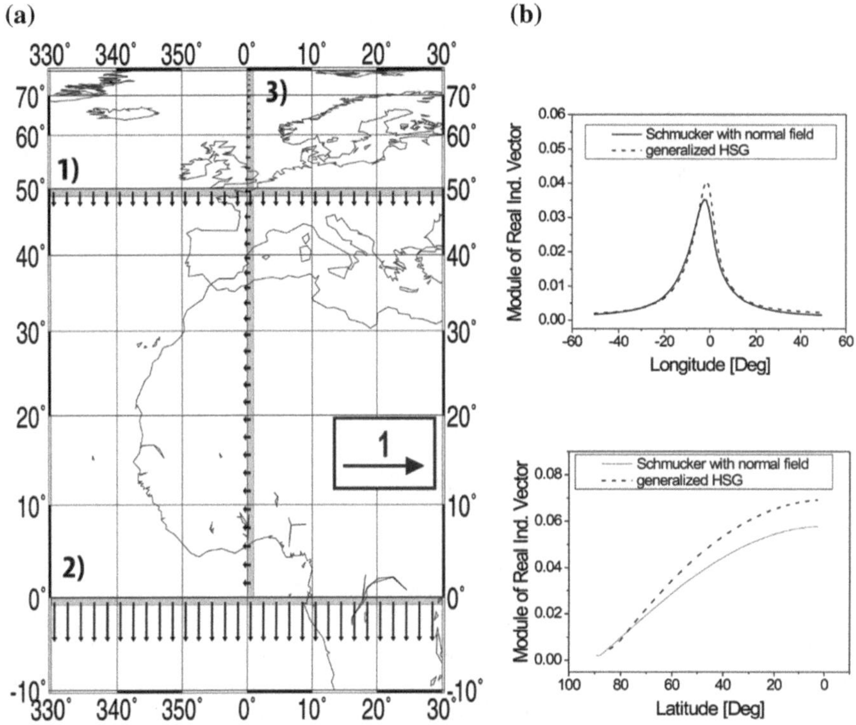

Fig. 3.4 The induction arrows of three different models: **a** the mantle step(*grey*) situated along the co-latitude 50°, **b** the step along the latitude 0°, **c** the step along longitude 0° at same depth as in Fig. 3.2

depend on the chosen depth stepposition measured in co-latitude (latitude), i.e., the anomaly "cap" spreads over the sphere as the co-latitude increases; see cases (a) and (b) in Fig. 3.4. A similar effect of reduction of the induction arrow size with decreasing co-latitude may be observed also for the case (c) when the anomaly is positioned along the meridian. Precise geophysical interpretation of this effect would require further modeling and is beyond the topic of this study.

The next model includes the 3D spherical mantle structure like a "diapir". This body has a resistivity of 10 Ω m and is situated at depths from 170 to 270 km with horizontal longitude dimension of 20° and taking from 30° to 50° in latitude (see the black-bordered sector in Fig. 3.5). The model was considered to be under the source field of the pure Dst variations with a period of 6 h. The calculated induced fields on the Earth's surface were used as input data for the numerical solving of the relation (3.1) by finite difference method. The grid of scalar impedances obtained by the MVS and GDS methods was converted into the apparent resistivities and to MVS gradient tippers which were then recalculated into induction arrows. The results obtained for GDS and MVS are presented in Fig. 3.5.

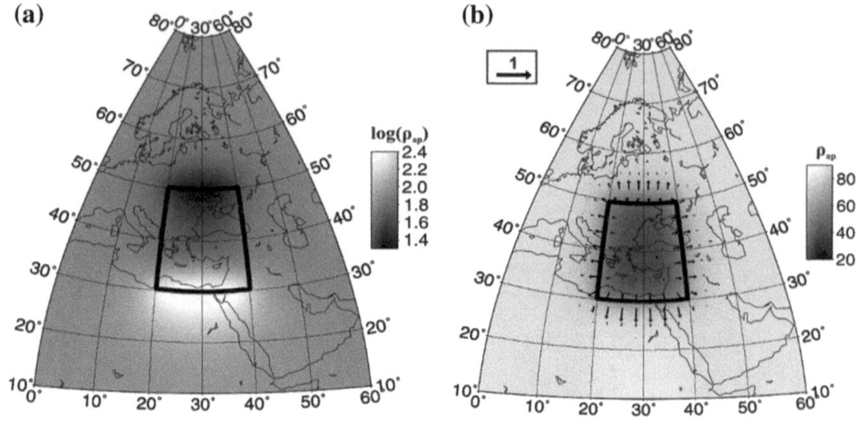

Fig. 3.5 The apparent resistivity anomaly above the 3D body produced by using the GDS method (*left*) and the MVS method (*right*) similar to the MTS method. The period is 6 h. Real induction *arrows* are shown in the right section

The left side of the picture shows that the GDS method is able to detect only two latitudinal boundaries of the anomaly sector in terms of the (anomalous) apparent resistivity. By contrast, the MVS method well reflects the influence of the whole anomalous body as well as the induction arrows. Note that the obtained result depends on the assumed polarization of the exciting field.

Finally, the model with mantle and surface inhomogeneities together was considered. This model includes the surface shell of the global inhomogeneous conductance [updated by Vozar et al. (2006)] and the layer depth stepin the upper mantle, as described in the first model. The MVS responses have been compared with the MTS ones. Originally, the MTS responses were in the form of main tensor impedances in the North-South (NS) and East-West (EW) directions (across and along the layer depth step, correspondingly) for the period of six hours and recalculated afterwards into apparent resistivities. The analyzed profile is situated along 20° meridian from the Arctic Ocean to the Mediterranean Sea; see Fig. 3.6.

It is clearly seen that the influence of the mantle inhomogeneity on the apparent resistivity distribution is weak (the calculated change in magnitudes is about 20 Ω m), i.e., it is orders of magnitude smaller than the background variations of the MTS apparent resistivities caused by the surface inhomogeneities. It is worth emphasizing that the MVS and MTS impedance phases in the WE direction (spherical E-polarization) have the same shapes just above the depth step of the upper mantle layer. So, the MTS phase values are disturbed much less by the surface inhomogeneities than the apparent resistivities themselves and their anomalous value (about 10°) could be observed on the background of the regional phase variations.

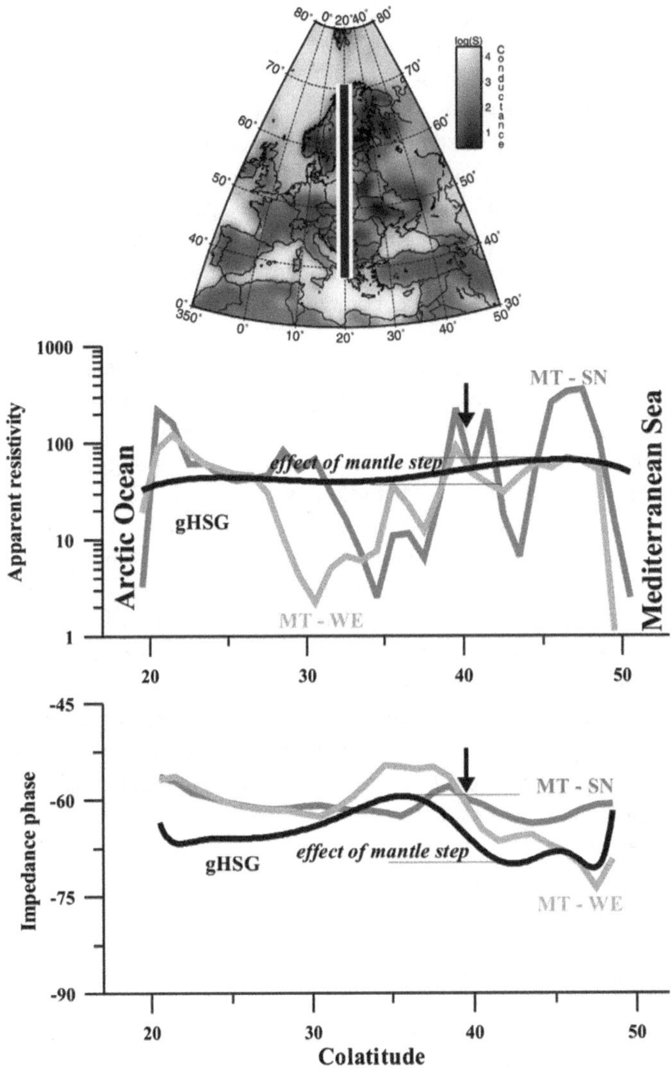

Fig. 3.6 Apparent resistivities (at the *top*) and impedance phases (at the *bottom*) modeled by the MTS soundings (*gray*) for two orthogonal directions, SN and WE (*gray*) and by MVS method (*black*). The profile is crossing Europe from the Arctic Ocean to the Mediterranean Sea along a longitude of 20°. The period is 6 h

3.4 Conclusions and Discussion

It was found that HSG and GDS methods can produce the response functions above the mantle inhomogeneities differing from both MVS and MTS ones (even free from the galvanic distortions). This fact is not much surprising if we bear in mind that those methods are just simplified modifications of the MVS for sounding the laterally homogeneous Earth. But the presented analysis performed on synthetic data showed that 1D inversion of the HSG and GDS responses as well as of their combinations with the responses obtained by MTS computed above the inhomogeneous media caused unrealistic results. The principal result of the modeling is the coincidence of responses above the mantle inhomogeneity obtained by MVS and MTS (without shift-effects) methods, both in the space and frequency domains. This consistence of the response functions obtained independently by MVS and MTS suggeststhe reliability of the developed approach. It also means that the impedances so determined can be combined for the joint inversion with more confidence.

It is a known fact that the static shift influences only the amplitudes of the resultant apparent resistivities, but not their phases. The described MVS method can give inconsistent results with the ones obtained by MTS when a shift-effect caused by subsurface inhomogeneities is taken into account, as suggested by Fig. 3.6. However, the effect of galvanic distortions is observed only for the modules of impedances (apparent resistivities), as expected, but not for their phases at long periods (depending on thickness and positioning of the inhomogeneity for a particular case). That is why merging and subsequent 1D inversion of the MVS and MTS phases alone is recommended on the grounds of this work. Besides, bias is mainly a characteristic of the estimations of amplitudes in contrast to phases in the linear relations. At this point it is worth to mention that this advantage of the phase fit can be omitted if the real and imaginary part of the *C* responses would be used instead of the apparent resistivity representation. In fact both, the real and imaginary part of *C* response, may undergo distortion caused by the galvanic effect.

Another problem of such a combination of methods is uncertain merging of the MVS scalar with MTS tensor responses even for their phase data. The performed numerical simulations have shown that phases of the MTS responses in the direction along the deep inhomogeneity (spherical E-mode) are much closer to the MVS ones as compared to the MTS responses perpendicular to the inhomogeneity border. The MVS responses parallel to the inhomogeneity border are the most sensitive to deep conductive anomalies. The most convenient choice of such MTS direction could be done by several ways in practice; see e.g. (Semenov et al. 2008). Note that the MVS responses are also slightly influenced by the surface inhomogeneities, manifesting themselves in their magnetic components of the computed field, too, in the period range of several hours (Everett et al. 2003). However, this influence is much weaker than that observed at the MTS ones (for the land-sea zones in Fig. 3.6).

The reliable information about the location of deep inhomogeneities can be obtained from the induction arrows calculated from the gradient tippers of the MVS method on the surface. Physically, they represent the real and imaginary parts of gradient of the scalar magnetovariation *C*-response function. As shown by the presented results, these induction arrows depend on latitudes (Figs. 3.4 and 3.5), which is related with the sphericity of the model. In practice, the information about the induction arrows can be obtained from the measured magnetic fields for the period range from hours up to a day (Schmucker 2003). At this period range, the Wiese–Parkinson relation (1.3) in Sect. 1.1 cannot be applied anymore, because its validity is constraint to the assumption of a homogeneous exciting field source only (Banks 1981).

References

Banks, R.J.: Strategies for improved global electromagnetic response estimates. J. Geomag. Geoelectr. **33**, 569–585 (1981)

Berdichevsky, M., Dmitriev, V.: Models and Methods of Magnetotelluric. Springer, Berlin Heidelberg (2008)

Everett, M.E., Constable, S., Constable, C.G.: Effects of near-surface conductance on global satellite induction responses. Geophys. J. Int. **153**, 277–286 (2003)

Guglielmi, A.V.: Hydromagnetic diagnostics and geoelectrical exploration. Uspehi Fizich. Nauk **152**(4), 605–637 (1989) (in Russian)

Kuvshinov, A., Utada, H., Avdeev, D., Koyama, T.: 3-D modeling and Analysis of the Dst EM responses in the North Pacific Ocean region. Geophys. J. Int. **160**, 505–526 (2005)

Schmucker, U.: Horizontal spatial gradient sounding and geomagnetic depth sounding in the period range of daily variation. In: Protokoll über das Kolloquium elektromagnetische Tiefenforschung ISSN 0946–7467 20.Konigstein, Kolloquium, pp. 228–237. 29.09–3.10 (2003)

Semenov, V. Yu., Ladanivskyy, B., Nowozynski, K.: Testing of the new HSG soundings in Central Europe, EGM 2007 International Workshop Innovation in EM, Grav. and Mag. Methods. Capri, Italy, April 15–18 (2007)

Semenov, VYu., Pek, J., Ádám, A., Jóźwiak, W., Ladanyvskyy, B., Logvinov, I.M., Pushkarev, P., Vozar, J.: Electrical structure of the upper mantle beneath Central Europe: Results of the CEMES project. Acta Geophys. **56**(4), 957–981 (2008)

Vozar, J., V. Yu. Semenov, A. Kuvshinov., C. Manoj.: A new subsurface map of the Earth conductance. Eos Trans. AGU **87**(33), pp. 326, 331 (2006)

Chapter 4
Results of Deep Soundings in Europe

Abstract In the first half of the XX century, there appeared new approaches to deep induction soundings. The theory of magnetovariation as well as magnetotelluric soundings was formulated just before the World War Two. Spatial derivatives of response functions (induction arrows) were obtained for the long periods. New phenomena have been detected by this method: secular variations of the Earth's apparent resistivity and the rapid changes of induction arrows over the last 50 years. The first ones can be correlated with the number of earthquakes and the second ones —with geomagnetic jerks in Central Europe. Extensive studies of geoelectrical structure of the crust and mantle were realized in the frame of a series of international projects. New information about geoelectrical structures of the crust in Northern Europe and Ukraine was obtained by deep electromagnetic soundings involving powerful, controlled sources. An influence of the crust magnetic permeability on the deep sounding results was confirmed.

Keywords Mantle · Conductivity · Apparent resistivity

4.1 Introduction

The second part of the XX century was characterized by an increasing interest in the deep induction soundings of the Earth mantle, particularly by the magnetovariation (MV) method (e.g., Roberts 1984, 1986; Schultz and Larsen 1987, 1990; Semenov 1989; Schultz 1990; Olsen 1992; Schultz et al. 1993; Olsen 1998, 1999a; Semenov and Jozwiak 1999; Schmucker 1999) and using satellite data (e.g., Oraevsky et al. 1993; Olsen 1999b). A feature of deep soundings is the lack of direct verification of the obtained results except of the ultra-deep borehole data in the Earth's crust, reaching 12 km depth. Therefore, the correctness of the induction sounding theory plays a dominant role in such investigations. For example, it is obvious that the model of the source field in the common form of a «plane wave» will not be valid

© Springer International Publishing AG 2018
V. Semenov and M. Petrishchev, *Induction Soundings of the Earth's Mantle*,
GeoPlanet: Earth and Planetary Sciences, DOI 10.1007/978-3-319-53795-5_4

for long periods as well as without considering the sphericity of the Earth (Schultz and Zhang 1994).

In retrospect, it is surprising that some oldest works in geoelectricity were broader than the traditional «plane wave» model. The concept of impedance in Russia was introduced in the early 1930s by Leontovich. Then his student Rytov has published the mathematical model (Rytov 1940), the first approximations of which are applied now as magnetotelluric (MT) and generalized magnetovariation (MV) soundings. Later on, a similar work was published by Wait (1954). Leontovich (1948) has considered limits of applicability of the Rytov model, and finally Senior and Volakis (1995) found a small error in that work. The simplified model suggested by Tikhonov (1950) and Cagniard (1953) certainly had a great success for the exploration of mineral resources by the MT method. Two traditional MV methods for estimation of impedances and tippers separately required separation of the observed field to "normal" and "anomalous" parts, which is a vague procedure. Solution of this problem has been already incorporated in the Rytov model, as pointed out by Guglielmi and Gokhberg (1987), while the simplified MV method was already developed and used by Banks (1969), Berdichevsky et al. (1969), Schmucker (1970). Approaches considered by Bates et al. (1976), Woods and Lilley (1979), Kuckes (1973) and Kuckes et al. (1985) were closer to Rytov's one. The corrected concept for the induction sounding impedances was proposed by Shuman (1999). The transformations, based on the Maxwell equations, of impedance matrix (called «tensor» with an overstatement because we do not know exactly what kinds of field sources form the measured signal in each direction) and an assumed resistivity azimuthal tensor was obtained theoretically by Reilly (see in: Weckmann et al. 2003), Semenov (1988, 2000). As a result, we have polar diagrams of the apparent resistivity instead of impedance ones. They can be not equal, as shown below.

To obtain reliable results, the deep induction soundings request impedances in a wide period range. To satisfy this request, the joint inversions of local MT impedances with regional MV responses were tested by Semenov (1988), Egbert and Booker (1992), Schultz et al. (1993), Semenov and Rodkin (1996). This approach was used for the induction soundings in the frame of the following international projects: "Baltic Electromagnetic Array Research" (BEAR 1998–2002), "Central Europe Mantle Electrical Structure" (CEMES 2001–2003) "Electro-Magnetic soundings of Trans-European Suture Zone" (EMTESZ 2003–2005), "Electro-Magnetic Mini Arrays" (EMMA 2005–2008) in Fennoscandia, "Fennoscandian Electrical conductivity from soundings with Natural and Controlled Sources" (FENICS 2007–2009) and "Lithospheric Structure of TESZ by Magneto-Variation Soundings" (LS-MVS 2009–2012) including territories of Belarus, Czech Republic, Finland, Germany, Hungary, Norway, Poland, Romania, Russia, Slovakia, Sweden and Ukraine. The national and global investigations have been already reviewed by Korja (2007), Kuvshinov (2012).

4.2 Soundings of 1D Layered Earth

The conductivities of the layered Earth's mantle were estimated using the magnetovariation method down to 2000 km depth. For this purpose, the responses obtained previously at seven geomagnetic observatories, IRT, KIV, MOS, NVS, HLP, WIT and NGK (Semenov 1998), were analyzed, together with reliable results published for harmonics of 11-year variations. Published models of the lower mantle conductivity obtained using the secular, 30–60 year. variations were also considered, in order to estimate the conductivity at depths down to the core. The new regional model of the lower mantle conductivity does not contradict modern induction sounding results. This model supports the idea that the mantle base, situated below 2000 km depth, has a very high conductivity.

The conductivity structure of the Earth's lower mantle has been estimated using both the electromagnetic induction and the internal source methods by McDonald (1957), Kolomyceva (1972), Alldredge (1977), Ducruix et al. (1980), Papitashvili et al. (1982); estimates have also been obtained from mineral physics by Shankland et al. (1993). However, "the picture of physics electric conductivity in the lower mantle is confusing" (Merrill et al. 1996). The conductance of the lower mantle estimated from 1000 km up to the core (2900 km), for example, varies by more than three orders of magnitude. Nevertheless, geophysicists suggest that a layer at the mantle base may have a very high conductivity (Zharkov 1983; Alldredge 1977). The conductivity of the 200–300 km thick D' layer at the mantle base might reach 100,000 S/m (Knittle and Jeanloz 1989), but the minimal value of the lower mantle conductivity has been estimated as 1 S/m (Coe et al. 1995). For this reason, we have attempted to use the induction method to as great a depth as possible, and then to compare with the results obtained by other authors.

Most of the regional estimations from the responses obtained by the induction soundings reveal the conductivity structure down to a depth of about 1000 km (Roberts 1986; Schultz 1990; Egbert and Booker 1992). This limitation is connected with the longest considered periods of the responses (max. 100 days). Using the monthly mean values of several selected geomagnetic observatories with high-quality observations over 22–43 years, the responses can be estimated from Dst variations over periods of up to two years (Semenov 1998). The mean responses up to periods of 700 days of seven mid-latitude geomagnetic observatories in the Eurasia region were chosen for interpretation. These data have been published together with details of their processing, including the remote reference and robust techniques, in Semenov (1998). In addition, these data have been combined with the mean daily responses for the European region and the global responses obtained for the harmonics of the 11-year variations by Yukutake (1965), Courtillot and Le Mouel (1976), Harwood and Malin (1977), Isicara (1977).

First of all, the forward spherical modelling was carried out to estimate the resolving power of the data composition considered. For this purpose, models of a weakly conductive earth (0.0001 S/m) with and without the conductive core (300,000 S/m) were constructed. The responses and the forward modelling curves

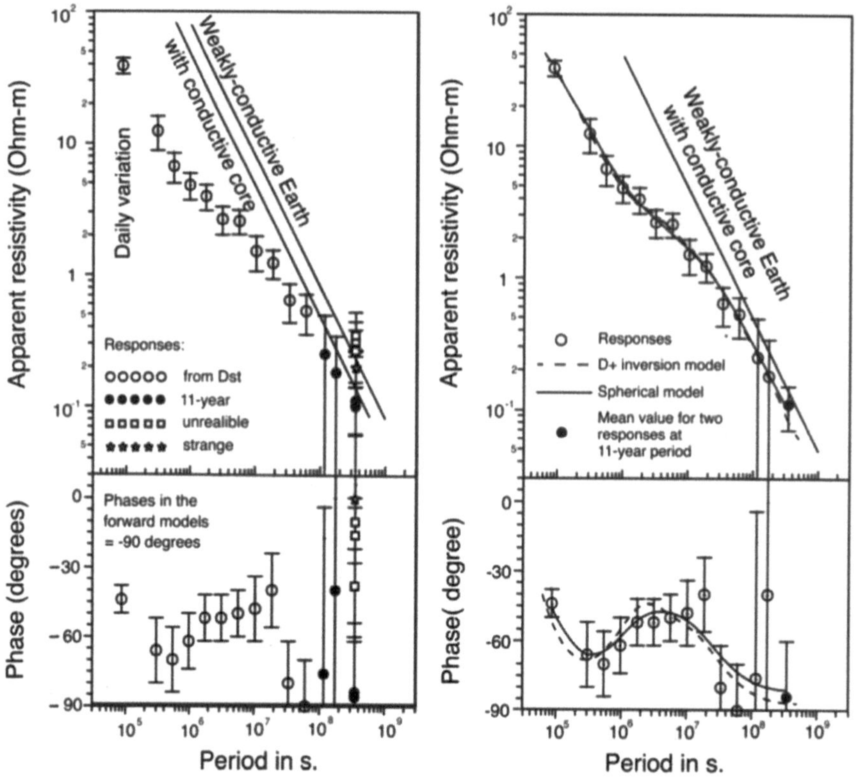

Fig. 4.1 The collected data and the forward spherical modelling responses for the weakly conductive Earth with and without the conductive core (*left*). Agreement of the experimental responses for 1-D inversions with the data that are not in contradiction with the induction method and with the existence of the conductive core (*right*). After Semenov and Jozwiak (1999)

obtained (lines) are shown in Fig. 4.1. For long periods, the forward modelling curves depend only on the spherical geometry (and not on the exact resistivity value of the weakly conductive earth), and the apparent resistivity phases are equal to −90° for these periods (Srivastava 1966; Sochelnikov 1979). This means that any experimental resistivities must be less than these resistivity values calculated for a weakly conductive earth (without the conductive core).

Most of the experimental data satisfy this condition, except for a few of the responses for the 11-year period (see Fig. 4.1, left). The fact that some responses show discrepancies with the assumption of the magnetovariation induction method (that is, the model of an external source) have already been noted (Rokityansky 1982). We will reject these unreliable responses from the data set. Thus, we conclude that most of the responses of the 11-year variations, estimated 20 years ago, are not very reliable, a fact which is reflected in the very large error bars for all their harmonics. If the reason for this is only the poor quality of the used data, due to the insufficient duration of observations for analyzing such long periods, future

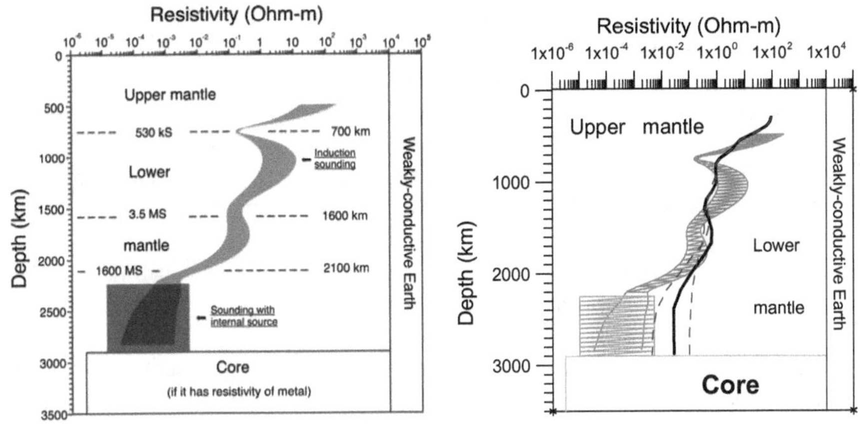

Fig. 4.2 The assumed layered model of the spherical Earth. Corresponding 1D inversion results of data presented in Fig. 4.1, taking into consideration data of the electro-jet 1969 year (*left*). The analogous result obtained by Olsen (1999) is shown (*right*) without consideration of the electro-jet data

experimental data could give more reliable information about the mantle conductivity.

For now, however, we selected only the few responses for our inversions that have no discrepancy with an external source. The selected data composition is shown in Fig. 4.2. One of the residual responses at the 11-year period is in contradiction [especially its phase, marked as strange in Fig. 4.1 (left)] with the above model of the Earth that includes a conductive core (if it is metallic). That is why we consider two data compositions: with and without the 11-year response. This can be done by considering the median or mean values as the averaged response for the 11-year period. However, the mean value including the unreliable response (which is not shown in Fig. 4.2) will be used only for an estimation of the minimal conductance of the lower mantle.

First, Parker's D+ inversion (Parker and Whaler 1981) was applied to the data composition that included the response that was in contradiction with the existence of the Earth's conductive core. The result was improved by using Weidelt transformation (Weidelt 1972) from plane to spherical geometry. The main conductive feature obtained with this inversion was detected at 2100 km depth. Its conductance is about 120 MS. In addition, conductive layers were detected at depths of 700 km (530 kS) and 1600 km (3.5 MS), as shown in Fig. 4.2. The 1-D OCCAM inversion (Constable et al. 1987) recalculated in a spherical geometry shows, in addition to the conductive zones detected by the D+ inversion, that the conductivity of the mantle base and the upper core is only about 10 S m^{-1} (for comparison, a metallic core must have a conductivity of about 100,000 S m^{-1}). For this reason we believe that this data composition is the least reliable. However, the value of 120 MS may be considered as a minimal conductance of the lower mantle (the maximum conductivity occurring mainly at 2100 km depth, if the CMB zone were even weakly conductive) estimated by this method.

The same algorithms were applied to the data composition without that unreliable response value. According to the D+ inversion, the mantle conductance at 2100 km depth reaches 1600 MS. The agreement between the calculated and experimental responses is shown in Fig. 4.1 (right). Next, the original spherical inversion (Jozwiak 1993) was used to analyze this data composition. The resistivity distribution (confirmed by the OCCAM inversion) is presented in Fig. 4.2. The mid-mantle conductive layer is clearly seen in this figure, together with the decreasing mantle resistivity: from 1000 to 2100 km.

Thus, both independent inversions have shown a similar geoelectrical structure of the mantle down to a depth of about 2100 km, beneath which a highly conductive layer with unknown thickness is required by the induction method. A model with a poorly conductive mantle base and highly conductive core seems to be unreliable, as a 'model with an abrupt jump of conductivity on the core–mantle boundary would be in contradiction with the observed data' (Kolomyceva 1972).

Now let us assume that the geoelectrical structure of the lower mantle obtained above is reliably estimated by the induction method up to 2100 km depth. We are going to compare and combine this result with other data obtained using the internal source variations with 30- and 60-year periods. Excluding the results obtained by McLeod (1994) and Ducruix et al. (1980), in which the lower mantle conductance is less than 120 MS (the minimal lower mantle conductance according to this study), we will estimate conductances from 1000 to 2900 km depth from other results.

By simple integration of the published conductivity distributions we have estimated that the lower mantle conductances vary from 150 MS (McDonald 1957) to 77,000 MS (Kolomyceva 1972). More credible values were suggested by Papitashvili et al. (1982): about 900–1600 MS (for comparison, the induction method also gave a value of 1600 MS). However, it follows from the above model that the lower mantle conductance is not greater than 5 MS down to 2000 km depth. We will therefore consider the above-calculated conductances relating only to the mantle base (from 2100 to 2900 km). The conductivity recalculated from the above-mentioned extreme conductances for this 800-km layer will vary from about 250–95,000 S m^{-1}, as shown for the resistivity values in Fig. 4.3. The agreement between the results obtained by the induction and internal source methods is good (see Fig. 4.2).

The model of the mantle conductivity presented here has been constructed using deep magnetovariation sounding data only. Generally speaking, the model is in agreement with the conductivity distributions obtained down to 1000 km depth for other regions (Schultz 1990). Below 2100 km, the model is not in contradiction with lower mantle conductivity estimates obtained by the internal source method.

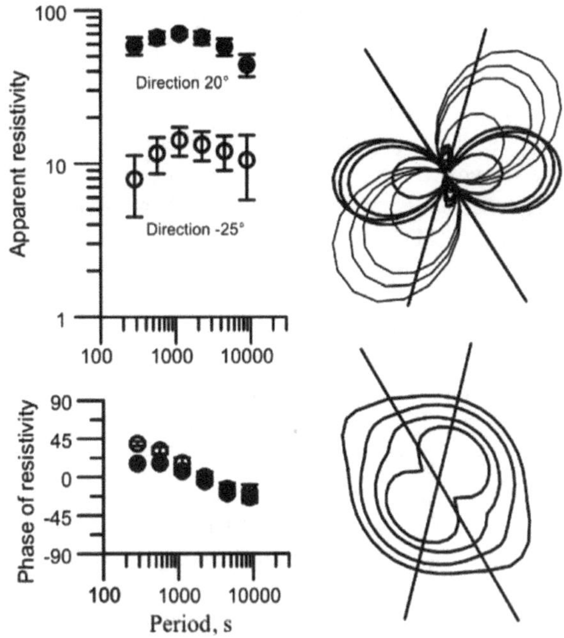

Fig. 4.3 Apparent resistivity and phase *curves* with appropriate polar diagrams. After Semenov et al. (2008)

4.3 Methods and Sources

Let us note that the magnetotelluric impedance $Z(\omega)$ for a fixed ω is a ratio $\boldsymbol{E}(\omega)/\boldsymbol{B}(\omega)$ because the measured magnetic field is the magnetic induction **B**(t), but not the field intensity **H**(t), as evidenced by its unit—nana Tesla (equal to 1 γ—off-system unit). Exactly this unit had been required by the IAGA in the resolution number 3 of «Transactions…» (1973). Note that the impedance at a fixed period is a *functional* of the conductivity (Berdichevsky and Zhdanov 1981). It means that many distributions of conductivity can correspond to the impedance value found for a fixed ω. This fact does not contradict the theorem of uniqueness of the inverse solution for induction soundings proved for *infinite input data* (Rokityansky 1982).

Estimations of impedances in practice are based on the random process theory assuming impedances as transfer functions between spectra of observed field components. These transfer functions are usually considered as scalar or matrix [2 × 2] values, sometimes even [3 × 3] (Dmitriev and Berdichevsky 2002). The last generalized heuristic approach leads to a similar result as obtained by Becken and Pedersen (2000). Such an approach is mixing both modes for the corresponding impedances that complicate an analysis of data over inhomogeneous media (Semenov and Shuman 2010). Separation of the modes for 3D case has been discussed by Becken et al. (2008). Note that the impedances can be found in the time domain too (Nowożyński 2004).

The publication by Weckmann et al. (2003) brings us back to the problem of transforming the MT impedance matrix to apparent resistivity tensor elements ρ_{ij}* for a laterally anisotropic medium. A widespread approach is considering the impedance as a tensor, two and sometimes four elements of which are recalculated into the same quantity of the scalar apparent resistivities even keeping indexes of the impedance tensor. Result of such procedure can look very strange. It is more natural considering *resistivity* of media as an *azimuthal tensor* and impedances as a matrix (for MT), vector (for GDS) or scalar (for GMV). The theoretical evidence transforming an impedance matrix to the resistivity tensor made by both Reilly (see in: Weckmann et al. 2003), Semenov (2000) are identical (see 1.9 in Sect. 1.3).

Of course, reconsidering long-standing postulates is a thankless job. But frequent recording of $\rho_{xy}* \sim Z_{xy}^2$ is without sense: asterisk xy marks a minor element of resistivity tensor while this asterisk marks the major element of impedance one. The relation following from the theory is $\rho_{xx}* \sim Z_{xy}^2 - Z_{xx} \cdot Z_{yy}$. To obtain a scalar resistivity, $\rho_{xx}* \sim Z_{xy}^2$ or $\rho_{yy}* \sim Z_{yx}^2$, the value of $Z_{xx} \cdot Z_{yy}$ must be minimized. Such directions were named the preferential ones; they may be not orthogonal (Fig. 4.3). The apparent resistivity modules shown here are only shifted at both preferential directions and their phases are similar (Fig. 4.3, right), while they are different for the orthogonal principal directions. Moreover, the directions of apparent resistivities obtained from the impedance matrix by common and theoretical transformation (Fig. 4.4) can be essentially different from their 1D inversion models.

A careful study of the anisotropic media with the arbitrarily directed tensor of conductivity has been made by Pek (2002), Pek and Santos (2002).

The problem of deep soundings arises: how an obtained local MT resistivity tensor can be combined with the regional GMV or Sq scalar resistivity or with the GDS continental vectoral one? In order to combine them, we have to assume that

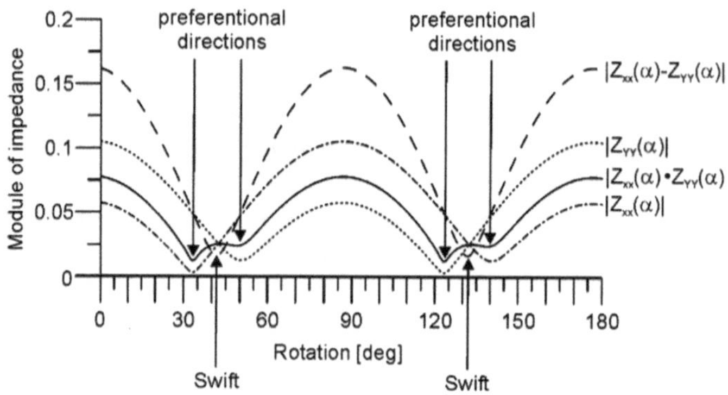

Fig. 4.4 Choice of two preferential directions (min $|Z_{xx} \cdot Z_{yy}|$) in comparison with the orthogonal principal directions (left) chosen by the Swift's procedure (min $|Z_{xx} - Z_{yy}|$). After Semenov et al. (2008)

the medium is rather homogeneous at great depths. But in fact the currents of different sources are induced and locked in different ways inside the Earth and may contain different information about its structure. This difference is clearly visible for the spherical model in Fig. 4.1 due to the different sources and methods including the Sq variations (Semenov et al. 2013). This problem still requires further efforts to study the deep irregularities in the mantle.

The problem of a lack of coincidence between inversion models of different teams in the international projects obtained from the same initial apparent resistivities is arising too. If it concerns 1D modeling, their results can be transformed to the monotonically increasing conductance with depth. This approach gives a possibility to investigate large areas presenting results as schemes of conductance at a fixed depth or depths to a fixed conductance, even without subsurface parts for comparison (Semenov and Jozwiak 2006).

Besides, it was established that secular variations of the Earth apparent resistivity estimated by the two methods, GMV and GDS, can reach 20% of the mean measured values (Fig. 4.5). These variations are well correlated with the number of earthquakes in the seismically active areas of Central Europe. The depths of the earthquakes are less than 40 km, in areas where sources of such variations are situated. So the internal as well as changing external sources can essentially disturb the sounding results during long observations. Besides, the induction arrows can change their directions and their values during a couple of years. This phenomenon coincides with the appearance of geomagnetic jerks in Central Europe (Petrishchev and Semenov 2013).

Another registered phenomenon is connected with the high magnetic permeability near the Curie point in the crust (Kiss et al. 2005). The influence of this effect on the deep soundings was considered by Szarka et al. (2007).

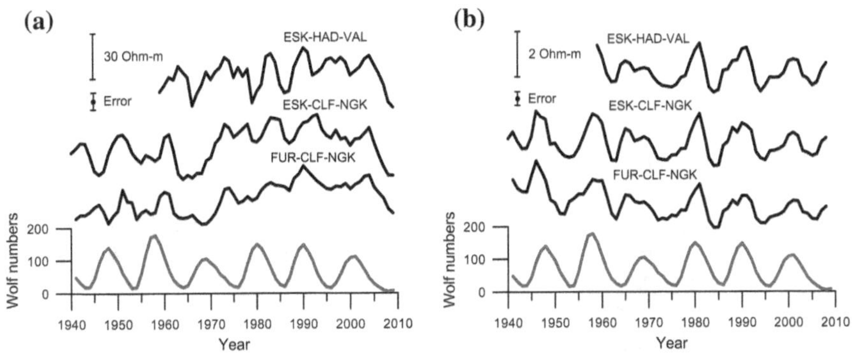

Fig. 4.5 Comparisons of the Earth's apparent resistivities with T = 8.8 h (a) and 30 days (b) observed by different groups of observatories marked by their codes. After Petrishchev and Semenov (2013)

4.4 Results of Deep Soundings

The final results of electrical conductivity studies of the Earth's crust at the Ukrainian and Fennoscandian shields were published by Ingerov et al. (1999) and Korja et al. (2002). Both studies show considerable heterogeneity in the crust conductance in these regions, reaching few orders of magnitude. The example of the crust conductance in the Ukraine is presented in Fig. 4.6.

The extensive deep geoelectrical studies in Northern Europe were realized in the frame of two international projects: BEAR (1998–2002) with its deep sounding continuation – EMMA (2005–2008) in Fennoscandia. These measurements have been done using c.a. 50 MT stations, displaced permanent layover the 150 × 150 km network on the territory of Fennoscandian shield. Four countries have taken part in the investigations. The obtained crustal conductance up to 60 km depth for Northern Europe is shown in Fig. 4.6 (right). The fixed conductance contrast reaches six orders and shows extremely high electrical heterogeneity of the crust in the Fennoscandian shield (Korja et al. 2002). The large regions of high resistance are surrounded by relatively narrow, highly conducting zones with conductances reaching dozens of kS. The origin of crustal anomalies can be connected with electronically conducting sulfide and carbon bearing structures (Zhamaletdinov 1996). The upper mantle conductance was estimated from the BEAR data during the EMMA project. It was shown that at depths of 150–300 km the conductance reaches 4–5 kS that may be expressed «as an asthenosphere conducting layer» (Sokolova and Varentsov 2007).

The second project was CEMES (2001–2003) at the territories of seven countries. The long-time MT measurements have been carried out at eleven geomagnetic observatories of Central Europe and their sounding results were combined with the MV soundings obtained at the same observatories using the historical hourly data (Fig. 4.7, left, top).

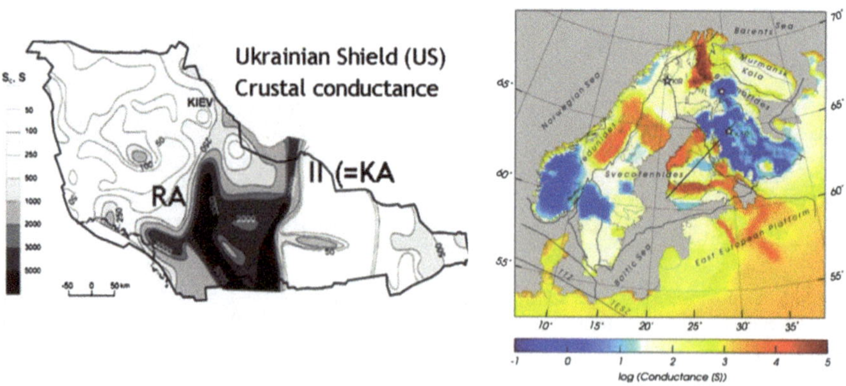

Fig. 4.6 Schemes of the crust conductances on the Ukrainian (*left*) and Fennoscandian (*right*) shields. After Ingerov et al. (1999), Korja et al. (2002), respectively

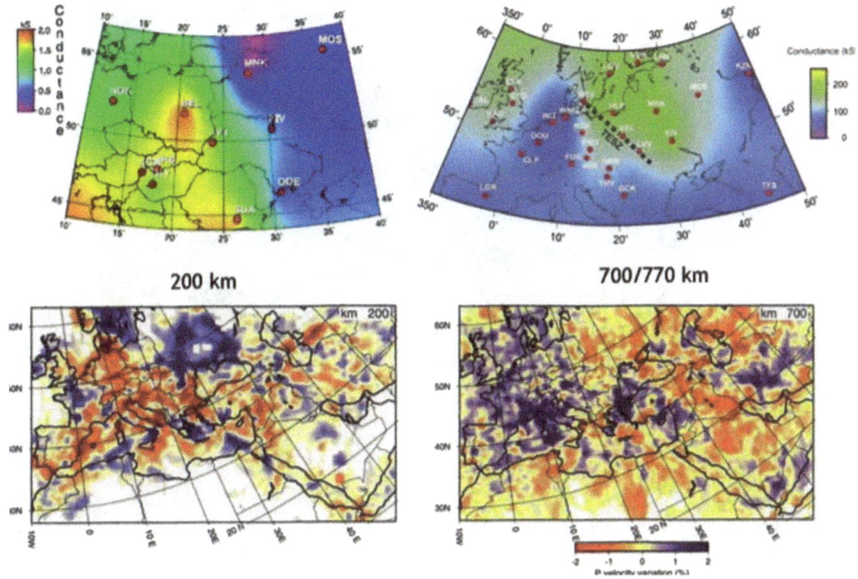

Fig. 4.7 Comparison of the conductance distributions in kS: depths 50–200 km (*left*, *top*) and depths 50–770 km (*right*, *top*) with seismic P-wave velocity variations (%) at 200 km (*left*, *bottom*) and 700 km (*right*, *bottom*). After Korja (2007)

Besides, the MV impedances for the periods from 4 h to 11 years estimated by six authors for 35 European observatories were collected. The precise selection of their results and subsequent combination allowed applying 1D inversion modeling to estimate the regional mantle conductance at a depth of 770 km beneath chosen observatories. The scheme of interpolation is shown in Fig. 4.7 (right, top).

These results have shown that the Trans-European Suture Zone (TESZ) coincides with the depth gradients of the 1 kS conductance in the upper mantle. The same effect has been observed along TESZ for the conductance gradient at a depth of 770 km in the mid-mantle. However, the conductance is increasing to the west in the upper mantle while in the mid-mantle it is increasing to the east. This reversing occurs by seismic data at about 600–800 km and has been confirmed by deep sounding results obtained along the Germany–Belarus profile (Fig. 4.8).

The international project EMTESZ (2003–2005) was carried out on the Polish and German territories. Two long magnetotelluric profiles along the seismic ones LT-7 (Guterch et al. 1994) and P2 (Janik et al. 2002) were rigorously studied by the MT method in the wide period range (0.1–10^4 s). The resistivity cross-sections up to upper mantle depths were published by Ernst et al. (2008).

Depths of the upper mantle were investigated separately combining three MT soundings in the TESZ—center of the profiles with apparent resistivities obtained at the nearest geomagnetic observatory Belsk situated in the TESZ too. The observed sounding result is a rarity: two curves of apparent resistivity in orthogonal

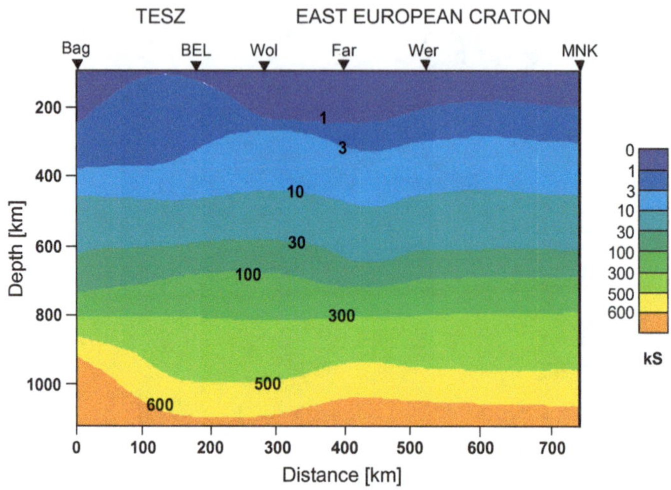

Fig. 4.8 Smoothed conductance deep structure along profile between geomagnetic observatories NGK (Germany) and MNK Belarus. After Semenov and Jozwiak (2005)

directions are reaching each other at the period of a day (Fig. 4.9, right). The obtained effect can be explained by the laterally anisotropic layer at depths of 30–40 km.

The next project was FENICS (2007–2009). Two mutually orthogonal industrial power transmission lines of 110 and 120 km lengths (Fig. 4.10) with the generator of 200 kW were used to estimate of the transversal resistance T ($\Omega{\cdot}m^2$) of the lithosphere in the frequency range of 0.1–200 Hz (Zhamaletdinov et al. 2011).

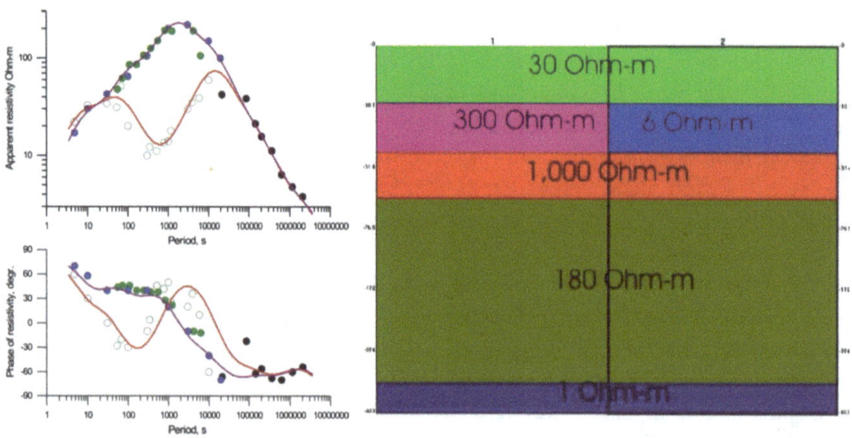

Fig. 4.9 The combined MT and MV apparent resistivities at two preferential directions crossing (*black*) and alongside (*red*) in the center of EMTESZ profiles (*left*) and both their 1D inversion models (*right*). After Semenov et al. (2005)

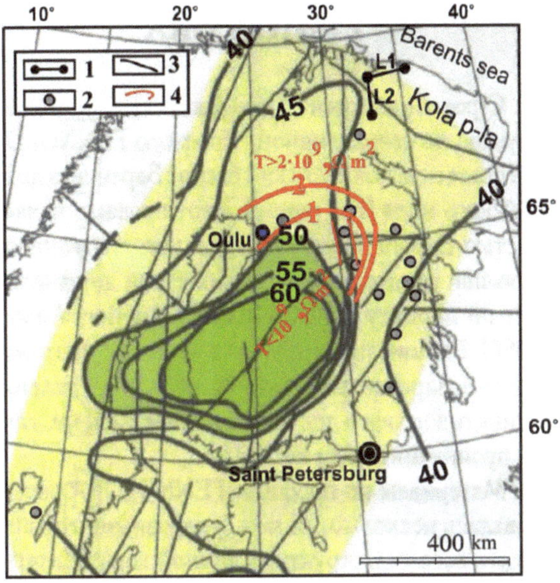

Fig. 4.10 The gradient of transversal resistance T ($\Omega \cdot m^2$) of the crust at the depth interval of 10–60 km (*red lines*) coinciding with the boundary of the deepest Moho zone in Europe (*black lines*). After (Zhamaletdinov 2011; Zhamaletdinov et al. 2011)

Spacing between transmitters and receiving points reached 700 km. Besides, the soundings with the Magneto-Hydro-Dynamic (MHD) generator «Khibiny» of 80 MW power were carried out for the investigations (Zhamaletdinov et al. 2005). The Moho depth gradient of the deepest anomalous zone in Europe (Grad and Tiira 2007) corresponds to the gradient of the transversal resistance.

Another interesting result of using the controlled sources was obtained in the western Siberia. The comparison of the ray logging data and deep induction soundings was carried out in the vicinity of a very deep (≈8 km) borehole. The combined sounding result is shown in Fig. 4.11. The combination of the apparent resistivities obtained with controlled and natural sources is commented in details by Zhamaletdinov et al. (2012).

The fifth international project was the LS-MVS (2009–2012). Five countries took part in these investigations. New MVS method was tested successfully in Central Europe using the data of geomagnetic observatories (Semenov et al. 2011). Thus, distributions of the induction arrows as well as response functions were estimated in Poland for the period range from 3 h to one day. This new information is analyzed now.

The original method to analyze the spatial distribution of induction arrows has been proposed recently by Jóźwiak (2012). Known arrows were recalculated to the scalar tippers and then into a ratio of the horizontal field at different points relative to a point at infinity. The Hilbert's transformation was used for this aim. These red zones are characterized by high conductance (Fig. 4.12). The obtained results are in good agreement with the geological knowledge.

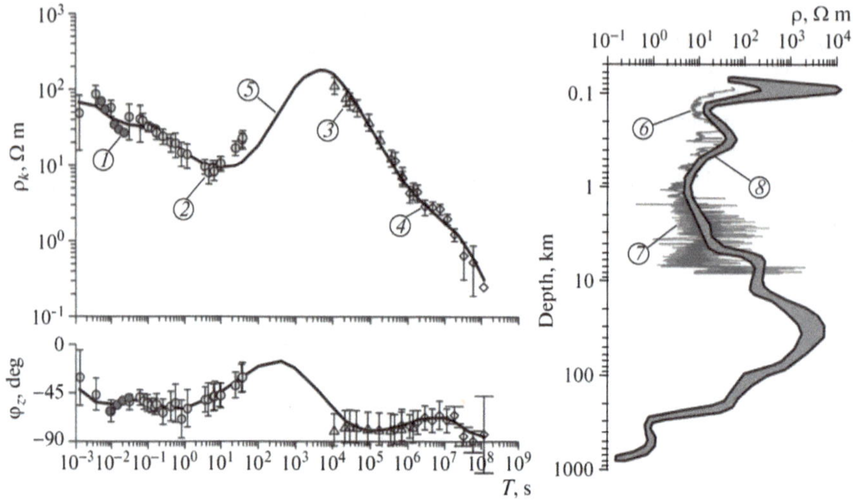

Fig. 4.11 The apparent resistivities and impedance phases obtained by the controlled and natural sources with: *1* ULF antenna, *2* MT data, *3*, *4* Sq and GDS method (observatory NVS), *5* 1D inversion, *6*, *7* ray logging results, *8* 1D inversion results. After Zhamaletdinov et al. (2012)

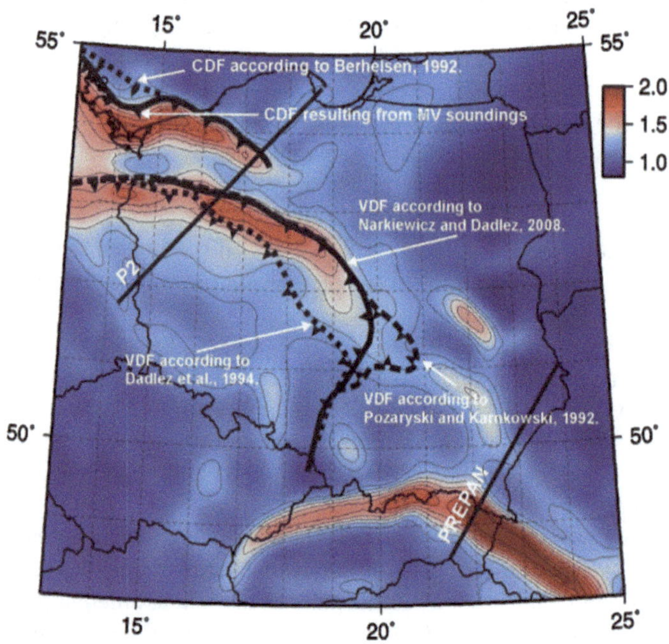

Fig. 4.12 Scheme of the spatial distribution of the conductive structures (*red*) basing on the hypothetical locations of the Caledonian and Variscan deformation fronts (*right*, scale in log of Siemens). After Jóźwiak (2012)

Recently, sounding results using data of the global geomagnetic observatory net were published by Praus et al. (2011). A review of electromagnetic study of lithospheric structure around the TESZ was made by Jóźwiak (2013).

4.5 Conclusions and Discussion

The development of generalized magnetovariation sounding theory significantly expanded possibilities of induction soundings in the period range from hours to a couple of days. The regional investigations by this method allow estimating additionally the gradients of response functions including information about electrical inhomogeneities in the upper mantle (up to ≈500 km) that was not made earlier. The induction sounding results are not stable: their long period variations can be caused by internal sources while variations with shorter periods by an abrupt change of the external field source. It is the reason why the results of formal inversions may be changeable in time. Besides, the forward spherical modeling has shown that sounding results using different sources and sounding methods above deep inhomogeneities can be essentially different. However, the sounding results of GMV and MT methods remain close to each other, even over significant irregularities in the Earth.

The methodic peculiarities of the mantle soundings are connected with combinations of two induction methods with tensor (MT) and scalar (GMV) or vectoral (GDS) apparent resistivities. The choice of directions in the MT soundings to compare with the MV ones is promising if the resistivity of medium is assumed to be a tensor rather than the impedance. It was shown that the conductance is most reliable for comparison between sounding results obtained by different investigators. An attractive but expensive method is that of controlled sources (like MGD generator) used in Russia for the deep soundings on a practically non-conductive surface. A lot of problems connected with natural sources disappear in such soundings, but requirements concerning the theory of the methods are not reduced.

Four of the five international projects in Central and Northern Europe performed areal studies. The results of the BEAR experiment increased the knowledge about the electrical structure of the Earth's crust at the Fennoscandian Shield (Korja et al. 2002). But another goal of the experiment—the search for a possible existence of the asthenosphere in the upper mantle—did not get a clear answer. Moreover, different research groups have come to some conflicting conclusions from the same experimental data. According to the St. Petersburg group, an intermediate conductive layer associated with partial melting of rocks is clearly recorded in the depth range of 200–400 km and its average value of the longitudinal conductance reaches 8 kS (Vardanyants and Kovtun 2009). This conclusion is not contrary to the interpretation of the seismic data which has a slightly different depth ≈100–150 km (Abramowitz et al. 2002). On the other hand, the interpretations made by other creative teams on the basis of the same experimental data assume the absence of any asthenosphere under the Fennoscandian shield or it is extremely weak on the background of the confidence limits (Varentsov et al. 2002; Sokolova and Varentsov 2007). The last interpretation of the BEAR data coincides with the interpretation made by Vanyan (2002). Discrepancies between the interpretations may be explained by sharp electrical inhomogeneity of the crust at this resistive region and high latitudes with complicated source field. It would be interesting to take into account the deep borehole in Karelia, where extremely solid rocks were met at the depth ≈12 km.

The two main results of the CEMES project have fixed gradients of the total conductance in the mantle at 300 and 770 km depths, coinciding with the TESZ. Now physical explanation of this effect is absent. It could be attributed to the methodological inaccuracies or errors of observations, but a similar inverse pattern was observed on the results of seismic tomography (Piromallo and Morally 2003). Reverses of the overabundant mass densities were also established at about 700 km depth by a rigorous analysis of the satellite data (Martinec and Pěč 1990). These phenomena require additional investigations as well as influence of the Earth magnetic permeability on the MV soundings and analysis of induction arrows at the period range of 3–30 h.

The FENICS project has allowed estimating the position of the transversal resistance gradients changing twice at the boundary of the deepest (≈50–60 km) area of the Moho in Northern Europe. These northern boundaries (gradient zones) of electrical and gravity anomalies coincide. Besides, the 1D inversions show the lower crust resistivity ($\approx 10^5\ \Omega \cdot m$), which is at least two orders of magnitude greater than for other shields in Europe and Canada. This high resistivity is coinciding well with the laboratory investigations for the 10–50 km depths (Zhamaletdinov et al. 2011) but is not consistent with previous studies (Vanyan et al. 2001) and other inversion results (Sokolova and Varentsov 2007).

References

Abramowitz, T., Thybo, H., Perchuc, E.: Tomographic inversion of seismic P- and S-wave velocities from the Baltic Shield based on FENNOLORA data. Tectonophysics **358**, 151–174 (2002)

Alldredge, L.R.: Deep mantle conductivity. J. Geophys. Res. **82**, 5427–5431 (1977)

Banks, R.J.: Geomagnetic variations and the electrical conductivity of the upper mantle. Geophys. J. R. Astron. Soc. **17**, 457–487 (1969)

Bates, R.H.T., Boerner, W.M., Dunlop, G.R.: An extended Rytov approximation and its significance for remote sensing and inverse scattering. Optics Commun. **18**(4), 421–423 (1976)

Becken, M., Pedersen, L.B.: Transformation of VLF anomaly maps into apparent resistivity and phase. Geophysics **68**(2), 497–505 (2000)

Becken, M., Ritter, O., Burckhardt, H.: Mode separation of magnetotelluric responses in the three-dimensional environments. Geophys. J. Int. **172**, 67–86 (2008)

Berdichevsky, M.N., Vanyan, L.L., Fainberg, E.B.: The frequency sounding of the Earth using spherical analysis results of geomagnetic variations. Geomag. Aeron. **9**, 372–374 (1969) (in Russian)

Berdichevsky, M.N., Zhdanov, M.S.: Interpretation of anomalies alternating electromagnetic field of the Earth. Nedra, Moscow (1981) (in Russian)

Cagniard, L.: Basic theory of the magneto-telluric method of geophysical prospecting. Geophysics **18**(3), 605–635 (1953). doi:10.1190/1.1437915

Coe, R., Prevot, M., Camps, P.: New evidence for extraordinary rapid change of the geomagnetic field during a reversal. Nature **374**, 687–697 (1995)

Courtillot, V., Le Mouel, J.L.: On the long-period variations of the Earth's magnetic field from 2 months to 20 years. J. Geophys. Res. **81**(N 17), 2941–2950 (1976)

Constable, S.C., Parker, R.L., Constable, C.G.: Occam's inversion: a practical algorithm for inversion of electromagnetic data. Geophysics **52**, 289–300 (1987)

Dmitriev, V.I., Berdichevsky, M.N.: A generalized model of impedance. Izv. Phys. Solid Earth **38** (10), 897–903 (2002)

Ducruix, J., Courtillot, V., Le Mouel, J.L.: The late 1960s secular variation impulse, the eleven year magnetic variation and the electrical conductivity of the deep mantle. Geophys. J. Roy. Astr. Soc. **61**, 73–94 (1980)

Egbert, G., Booker, J.R.: Very long period magnetotellurics at Tucson observatory: implications for mantle conductivity. J. Geophys. Res. **97**(B11), 15099–15112 (1992)

Ernst, T., Brasse, H., Cherv, V., Hoffmann, N., Jankowski, J., Jozwiak, W., Kreutzmann, A., Neska, A., Palshin, N., Pedersen, L.B., Smirnov, M., Sokolova, E., Varentsov, I.M.: Electromagnetic images of the deep structure of the trans-European Suture Zone beneath Polish Pomerania. Geophys. Res. Lett. **35**, L15307 (2008). doi:10.1029/2008GL034610

Grad, M., Tiira, T.: The Moho Depth of the European Plate. European Seismological Commission, Warsaw/Helsinki (2007)

Guglielmi, A.V., Gokhberg, M.B.: On the magnetotelluric sounding in the seismically active areas. Izv. Physics of the Solid Earth **33**(11), 122–123 (1987) (in Russian)

Guterch, A., Grad, M., Janik, T., Materzok, R., Luosto, U., Yliniemi, J., Luck, E., Schulze, A., Forste, K.: Crustal structure of the transition zone between Precambrian and Variscan Europe from new seismic data along LT7 profile (NW Poland and eastern Germany). Geophy. C.R. Acad. Sci. Paris t. **319**(Series II), 1489–1496 (1994)

Harwood, J.M., Malin, S.R.: Sunspot cycle influence on the geomagnetic field. Geoph. J. Roy. Astron. Soc. **50**, 605–618 (1977)

http://www.seismo.helsinki.fi/mohomap/; http://www.igf.fuw.edu.pl/mohomap/

Isicara, A.M.: Solar cycle geomagnetic variation. ActaGeodaet. Geophys. et Montan. Acad. Sci. Hung. **12**(1–3), 397-405 (1977)

Ingerov, A.I., RokityanskyI, I., Tregubenko, V.I.: Forty years of MTS studies in the Ukraine. Earth Planet Space **51**, 1127–1133 (1999)

Janik, T., Yliniemi, J., Grad, M., Thybo, H., Tiira, T., POLONAISE P2 Working Group 1.: Crustal structure across the TESZ along POLONAISE'97 seismic profile P2 in NW Poland. Tectonophysics **360**(1–4), 129–152 (2002). doi:10.1016/S0040-1951(02)00353-0

Jozwiak, W.: Application of a stochastic method to the global inverse problem. ActaGeophys. Polon. **41**(N 4), 523–533 (1993)

Jóżwiak, W.: Large-scale crustal conductivity in Central Europe and its correlation to deep tectonic structures. Pure. appl. Geophys. **169**, 1737–1747 (2012)

Jóżwiak, W.: Electromagnetic study of lithospheric structure in the marginal zone of East European Craton in NW Poland. Acta Geophys. **61**(4), 567–574 (2013)

Kiss, J., Szarka, L., Pracser, E.: Second order magnetic phase transition in the Earth. Geophys. Res. Lett. **32**, L24310 (2005)

Knittle, E., Jeanloz, R.: Simulating the core-mantle boundary: an experimental study of high-pressure reactions between silicates and liquid iron. Geophys. Res. Lett. **16**, 609–612 (1989)

Korja, T.: How is the European lithosphere imaged by magnetotelluric? Surv. Geophys. **28**, 239–272 (2007)

Korja, T., Engels, M., Zhamaletdinov, A.A., Kovtun, A.A., Palshin, N.A., The BEAR Working Group.: Crustal conductivity in Fennoscandia—a compilation of a database on crustal conductance in the Fennoscandian Shield. Earth Planets Space **54**, 535–558 (2002)

Kolomyceva, G.I.: On electrical conductivity distribution in the Earth's mantle from data of the secular variations of the geomagnetic field. Geomag. Aeron. **12**(6), 1082–1085 (1972) (in Russian)

Kuckes, A.F.: Relations between electrical conductivity of a mantle and fluctuating magnetic fields. Geophys. J. R. astron. Soc. **32**, 319–331 (1973)

Kuckes, A.F., Nekut, A.G., Thompson, B.G.: A geomagnetic scattering theory for evaluation of the Earth structure. Geophys. J. R. astron. Soc. **8**, 319–330 (1985)

Kuvshinov, A.: Deep electromagnetic studies from land, sea, and space. Progress status in the past 10 years. SurvGeophys. **33**, 169–209 (2012). doi:10.1007/s10712-011-9118-2

Leontovich, M.A. On approximate boundary conditions for an electromagnetic field on the surface of highly conductive bodies. In: "Issledovania po rasprostraneniu radiovoln", pp. 5–12. Academy of Sciences of the USSR, Moscow (1948) (in Russian)

Martinec, Z., Pěč, K.: The influence of the core-mantle boundary irregularities on the mass density distribution inside the Earth. In: A. Vogel, C.O. Ofoegbu, R. Gorenflo, and B. Ursin (eds.), *Geophysical Data Inversion. Methods and Applications, Proc. 7th Int. Math. Geophys. Seminar, 8–11 February 1989, Free University of Berlin*, 233–256 (1990). doi:10.1007/978-3-322-89416-8_15

McDonald, K.L.: Penetration of the geomagnetic secular field through a mantle with variable conductivity. J. Geophys. Res. **62**(1), 117–141 (1957)

McLeod, M.: Magnetospheric and ionospheric signals in magnetic observatory monthly means: electrical conductivity of the deep mantle. J. Geophys. Yes. **99**, 13577–13590 (1994)

Merrill, R.T., McElhinny, M.W., McFadden, P.L.: The magnetic field of the Earth: paleomagnetism, the core, and the deep mantle, p. 531. Academic Press, San Diego, California (1996)

Nowożyński, K.: Estimation of magnetotelluric transfer functions in the time domain over a wide frequency band. Geophys. J. Int. **158**, 32–41 (2004)

Olsen, N.: Day-to-day C-response estimation for Sq from 1 cpd to 6 cpd using the Z: Y method. J. Geomag. Geoelectr. **44**, 433–447 (1992)

Olsen, N.: The electrical conductivity of the mantle beneath Europe derived from C-responses from 3 to 720 hr. Geophys. J. Int. **133**, 298–308 (1998)

Olsen, N.: Long-period (30 days–1 year) electromagnetic sounding and the electrical conductivity of the lower mantle beneath Europe. Geophys. J. Int. **138**(179), 187 (1999a)

Olsen, N.: Induction studies with satellite data. Surv. Geophys. **20**, 309–340 (1999b)

Oraevsky, V.N., Rotanova, N.M., Bondar, T.N., Abramova, D.Y., Semenov, V.Y.: On the radial geoelectrical structure of the mid-mantle from magnetovariation sounding using MAGSAT data. J. Geomagn. Geoelectr. **45**, 1415–1423 (1993)

Papitashvili, N.E., Rotanova, N.M., Fishman, V.M.: Estimation of conductivity of the lower mantle from the analysis of 60- and 30-year variations of geomagnetic field. Geomagnet. Aerono. **22**(6), 1010–1015 (1982) (in Russian)

Parker, R.L., Whaler, K.A.: Numerical method for establishing solutions to the inverse problem of electromagnetic induction. J. Geophys. Res. **86**, 9574–9584 (1981)

Pek, J.: Spectral magnetotelluric impedances for an anisotropic layered conductor. ActaGeophys. Polonica. **50**(4), 619–643 (2002)

Pek, J., Santos, X.: Magnetotelluric impedances and parametric sensitivities for 1-D anisotropic layered media. Computers Geosciences **28**, 939–950 (2002)

Petrishchev, M.S., Semenov, V.Y.: Secular variations of the Earth's apparent resistivity. Earth Planet. Sci. Lett. **361**, 1–6 (2013)

Piromallo, C., Morelli, A.: P wave tomography of the mantle under the Alpine-Mediterranean area. J. Geophys. Res. **108**, B2, 2065 (2003). doi:10.1029/2002JB001757

Praus, O., Pecova, J., Cerv, V., Kovacikova, S., Pek, J., Velimsky, J.: Electrical conductivity at mid-mantle depths estimated from the data of Sq and long period geomagnetic variations. Stud. Geophys. Geod. **55**, 241–264 (2011)

Roberts, R.G.: The long period electromagnetic response of the Earth. Geophys. J. R. Astron Soc. **2**, 547–572 (1984)
Roberts, R.G.: The deep electrical structure of the Earth. Geophys. Roy. Astr. Soc. **85**(3), 563–600 (1986)
Rokityansky, I.I.: Geomagnetic investigation of the Earth's crust and mantle, p. 381. Springer, Berlin (1982)
Rytov, S.M.: Skin-effect calculations by the disturbance method. J. Exp. Theor. Phys. **10**(2), 180–189 (1940) (in Russian)
Schmucker, U.: Anomalies of geomagnetic variations in the southwestern United States. Bull. Scripps Inst. Ocean. **13**, 1–165 (1970)
Schmucker, U.: A spherical harmonic analysis of solar daily variations in the years 1964-1965: response estimates and source fields for global induction—I/II. Methods/Results. Geophys. J. Int. **136**, 439–476 (1999)
Schultz, A., Larsen, J.C.: On the electrical conductivity of the mid-mantle: I. Calculation of equivalent scalar magnetotelluric response function. Geophys. J. R. Astron. Soc. **88**, 733–761 (1987)
Schultz, A., Larsen, J.C.: On the electrical conductivity of the mid-mantle: II. Delineation of heterogeneity by application of extreme inverse solutions. Geophys. J. Int. **101**, 565–589 (1990)
Schultz, A.: On the vertical gradient and associated heterogeneity in mantle electrical conductivity. Phys. Earth Planet. Int. **64**, 68–86 (1990)
Schultz, A., Kurtz, R.D., Chave, A.D., Jones, A.G.: Conductivity discontinuities in the upper mantle beneath a stable Craton. Geophys. Res. Lett. **20**(24), 2941–2944 (1993)
Schultz, A., Zhang, T.S.: Regularized spherical harmonic analysis and the 3D electromagnetic response of the Earth. Geophys. J. Int. **116**, 141–156 (1994)
Semenov, V.Y.: Evaluation of mantle conductivity beneath northern hemisphere continents. Izv. Phys. Solid Earth **25**(3), 221–226 (1989) (in Russian)
Semenov, V.Y.: Analysis of magnetotelluric data during the anisotropic media sounding. Geology Geophys. **10**, 121–125 (1993) (in Russian)
Semenov, V.Y.: Regional conductivity structures of the Earth's mantle. Publications of the Institute of Geophysics, Polish Academy of Sciences, vol. C-65, issue no. (302), 122 pp (1998)
Semenov, V.Y.: On the apparent resistivity in magnetotelluric sounding. Izv. Phys. Solid Earth **36** (1), 99–100 (2000)
Semenov, V.Y., Jozwiak, W.: Model of the geoelectrical structure of the mid- and lower mantle in the Europe-Asia region. Geophys. J. Int. **138**, 549–552 (1999)
Semenov, V.Y., Jozwiak, W.: Estimation of the upper mantle electric conductance at the polish margin of the East European Platform. Izv. Phys. Solid Earth **41**(4), 80–87 (2005)
Semenov, V.Y., Jóżwiak, W.: Lateral variations of the mid-mantle conductance beneath Europe. Tectonophysics **416**, 279–288 (2006)
Semenov, V.Y., Ladanivskyy, B.T., Nowożyński, K.: New induction sounding tested in Central Europe. Acta Geophys. **59**(5), 815–832 (2011)
Semenov, V.Y., Ernst, T., Nowożyński, K., Pek, J., EMTESZ WG.: Estimation of the deep geoelectrical structure beneath TESZ in NW Poland. Publications of the Institute of Geophysics, Polish Academy of Sciences, vol. C-95, issue no. (386): Study of geological structures containing well-conductive complexes in Poland, pp. 63–65 (2005)
Semenov, V.Y., Rodkin, M.V.: Conductivity structure of the upper mantle in an active subduction zone. J. Geodyn. **21**(4), 355–364 (1996)
Semenov, V.Y., Shuman, V.N.: Impedances for the deep electromagnetic soundings. Acta Geophys. **58**(4), 527–542 (2010)
Semenov, V.Y., Ádám, A., Jóźwiak, W., Ladanyvskyy, B., Logvinov, I.M., Pek, J., Pushkarev, P., Vozar, J., Experimental Team of CEMES: Electrical structure of the upper mantle beneath Central Europe: Results of the CEMES project. Acta Geophys. **56**(4), 957–981 (2008)
Semenov, V.Y., Hvozdara, M., Vozar, J.: Modeling of deep magnetovariation soundings on the rotating Earth. Acta Geophys. **61**(2), 264–280 (2013)

Senior, T.B.A., Volakis, J.L.: Approximate boundary conditions in electromagnetics, p. 353. IEE Press, London (1995)
Shuman, V.N.: Scalar local impedance conditions and the impedance tensor in processing and interpretation of a magnetotelluric experiment. Geoph. J. Kiev. **19**, 361–385 (1999)
Shankland, T.J., Peyronneau, J., Poirier, J.-P.: Electrical conductivity of the Earth's lower mantle. Nature **366**, 453–455 (1993)
Sochelnikov, V.V.: Principles of the theory of the natural electromagnetic field in a sea. Gidrometeoizdat, Leningrad (1979) (in Russian)
Sokolova, E.Y., Varentsov, I.M., BEAR WorkingGroup.: Deep array electromagnetic sounding on the Baltic shield: external excitation model and implications for upper mantle conductivity studies. Tectonophysics **445**, 3–25 (2007)
Srivastava, S.P.: Theory of the magnetotelluric method for a spherical conductor. Geoph. J. Roy. Astr. Soc. **11**, 373–387 (1966)
Szarka, L., Franke, A., Prácser, E., Kiss, J.: Hypothetical mid-crustal models of second-order magnetic phase transition. 4th International Symposium on Three-Dimensional Electromagnetics Freiberg, Germany, September, pp. 27–30 (2007)
Tikhonov, A.N.: Determination of the electrical characteristics of the deep layers of the Earth's crust. Dokl. AN USSR. 73, 2, 295–297 (1950) (in Russian). Transactions of the second general scientific assembly, IAGA Bull., 35. Kyoto, Japan, 1973, p. 189
Transaction…: Transactions of the 2nd IAGA Scientific Assembly, Kyoto, Japan. IAGA Bull. **35**, 189 pp (1973)
Vanyan, L., Tezkan, B., Palshin, N.: Low electrical resistivity and seismic velocity at the base of the upper crust as indicator of rheologically weak layer. Surv. Geophys. **22**, 2, 131–154, (2001). doi:10.1023/A:1012937410685
Vanyan, L.L., Kuznetsov, V.A., Lyubetskaya, T.V., Palshin, N.A., Korja, T., Lahti, I., BEAR Working Group.: Electrical conductivity of the crust beneath Central Lapland. Izv. Phys. Solid Earth **38**(10), 798–815 (2002)
Vardanyants, I.L., Kovtun, A.A.: The study of the possible existence of asthenosphere on the territory of Fennoscandian shield by the BEAR data. In the book: Complex Geological-Geophysical Models of Ancient Shields. Apatity. Edition of the Geological Institute of the Kola Science Centre of Russian Academy of Sciences, pp. 15–18 (2009)
Varentsov, I. M., Engels, M., Korja, T., Smirnov, M.Y, The BEAR WG.: The generalized geoelectrical model of Fennoscandia: a challenging database for long period 3D modeling studies within Baltic electromagnetic array research (BEAR). Izv. Phys. Solid Earth **10**, 64–105 (2002) (in Russian)
Wait, J.R.: On the relation between telluric currents and the Earth's magnetic field. Geophysics **19**, 281–289 (1954)
Weidelt, P.: The inverse problem of geomagnetic induction. Geophysik. **38**, 257–289 (1972)
Weckmann, U.O., Ritter, O., Haak, V.: Images of the magnetotelluric apparent resistivity tensor. Geophys. J. Int. **155**, 456–468 (2003). doi:10.1046/j.1365-246X.2003.02062.x
Woods, D.V., Lilley, F.E.M.: Geomagnetic induction in Central Australia. J. Geomagn. Geoelect. **31**, 449–458 (1979)
Yukutake, T.: The solar cycle contribution to the secular change in the geomagnetic field. J. Geomag. Geoelectr. **17**, 287–309 (1965)
Zhamaletdinov, A.A.: Graphite in the Earth's crust and electrical conductivity anomalies. Izv. Phys. Solid Earth **32**(4), 272–288 (1996)
Zhamaletdinov, A.A.: Khibiny MHD experiment: the 30th anniversary. Izv. Phys. Solid Earth **41** (9), 737–742 (2005)
Zhamaletdinov, A.A.: The new data on the structure of the continental crust based on the results of electromagnetic sounding with the use of powerful controlled sources. Dokl. Earth Sci. **438** (Part 2), 798–802 (2011) (in Russian)
Zhamaletdinov, A.A., Shevtsov, A.N., Korotkova, T.G., Kopytenko, Y.A., Izmailov, V.S., Petrishchev, M.S., Efimov, B.V., Barannik, M.B., Kolobov, V.V., Prokopchuk, P.I., Smirnov, M.Y., Vagin, S.A., Pertel, M.I., Tereshchenko, E.D., Vasil'ev, A.N., Grigoryev, V.F.,

Gokhberg, M.B., Trofimchik, V.I., Yampolsky, Y.M., Koloskov, A.V., Fedorov, A.V., Korja, T.: Deep electromagnetic sounding of the lithosphere in the Eastern Baltic (Fennoscandian) shield with high power controlled sources and industrial power transmission lines (FENICS experiment). Izv. Phys. Solid Earth **47**(1), 2–22 (2011)

Zhamaletdinov, A.A., Petrishchev, M.S., Shevtsov, A.N., Kolobov, V.V., Selivanov, V.N., Esipko, O.A., Kopytenko, E.A., Grigorijev, V.F.: Electromagnetic sounding of the Earth's crust in the vicinities of the SG-6 and SG-7 super-deep boreholes in the fields of natural and powerful controlled sources. Dokl. Earth Sci. **445 1**, 889–893 (2012)

Zharkov, V.N.: Internal Structures of the Earth and Planets. Nauka, Moscow (1983) (in Russian)

Chapter 5
Electromagnetic Monitoring

Abstract This chapter is dedicated to electromagnetic monitoring by magneto-variation methods and reveals a possible connection of geoelectric characteristics with solar activity, seismicity and a very interesting phenomenon of geomagnetic jerk. It highlights the peculiarities of the data processing with examples; the technique to obtain stable and reliable results is shown. A method is presented for experimental detection of the direction to the geomagnetic pole that can be used for geomagnetic deep soundings. The presented successful approach to electromagnetic monitoring has allowed establishing a strong correlation with the solar activity and quasi-linear trend in the apparent resistivity that may indicate the changes of mantle conductivity. It is shown that there are several zones in Eurasia where the variations of apparent resistivity are related to integral seismicity with magnitude greater than 3. Some features in the induction vectors and apparent resistivity have been detected which coincide in time with geomagnetic jerks registered on the European geomagnetic network. A possible mechanism of geomagnetic jerk's influence on the induction sounding results is presented.

Keywords Conductivity changes · Regional seismicity · Geomagnetic jerk · Thermal lithosphere

5.1 Introduction

The electromagnetic induction soundings are based on the currents induced by some external fields in the Earth's interior. These fields are even able to penetrate the lower mantle and they contain information about the Earth's conductance. Commonly, the sounding results are considered to be independent of the variability of its source power due to the use of impedance boundary conditions. But it does not mean that they must be stable in time. Impedances can depend on other properties of external sources or media, such as magnetic permeability or porosity which change in time with pressure and temperature.

© Springer International Publishing AG 2018

V. Semenov and M. Petrishchev, *Induction Soundings of the Earth's Mantle*,
GeoPlanet: Earth and Planetary Sciences, DOI 10.1007/978-3-319-53795-5_5

Relations between variability of the electromagnetic fields and seismicity in some Earth's regions have been known for a long time (e.g., Tzanis 2010; Duma and Ruzhin 2003). Variability of electromagnetic field caused by piezoelectric and seismo-electro-kinetic phenomena inside the Earth was also known and successfully applied in exploration geophysics by several types of methods having the theoretical background (e.g., Guglielmi and Levshenko 1994; Svetov and Gubatenko 1999; Neustadt et al. 2006).

The search for earthquake or volcanic precursors is based on the apparent resistivity monitoring by the relatively high frequency magnetotelluric (MT) method in seismically active areas (e.g., Park et al. 1993; Lu et al. 1999; Ernst et al. 1993). The tidal variations of apparent resistivities were established by this very method (e.g., Zhamaletdinov et al. 2000; Saraev et al. 2010). However, the MT method is not reliable enough for the long periods on a land, because of decreasing signal-to-noise ratio in the electric field with increasing periods.

The modern magnetovariation methods are the generalized magnetovariation method (GMV) and the geomagnetic depth soundings (GDS). The main difference between these methods is the number of uncorrelated input signals: three from auroral source generating a quasi-plane wave with vertical component (Vanyan et al. 2002) for shorter periods (1.17), and one from the polarized *Dst* variations (1.7). The GMV algorithm is used for analysis of hourly data when the term $divB_\tau$ is most noisy (Semenov et al. 2011). The GDS algorithm (1.7) can be applied to the daily data when B_r is the noisiest component (Banks 1969). These criteria are necessary for selecting the most reliable *C*-response between the various estimations.

These methods are more suitable for investigating the temporal changes of the mantle apparent resistivity, especially for low frequencies (long periods), and their relation to the different sources, as we will try to demonstrate in this chapter.

5.2 Data Processing Peculiarities

As described above, the sounding results depend on many factors, such as changes of source position or its configuration, magnetic permeability, seismicity and so on. When such changes occur, it is possible to obtain something like that presented in Fig. 5.1 for changes of induction arrows in magnetic storm time. Meanwhile, the soundings mean that the Earth's resistivity is stable in time. One of the possible ways to obtain more-or-less stable results is to average the responses for a selected time. In this case, we will lose the time sensitivity but increase the stability of results.

The coherence criteria can be applied to select the reliable results at each period among the series of each data segment. For example, *C*-responses can be considered valid if the coherences between output B_r and input $divB_\tau$ signals are greater than 0.7 for GMV and GDS methods. In this case, the mean partial coherences between the output B_r and input B_υ and B_φ signals should be greater than 0.3. The

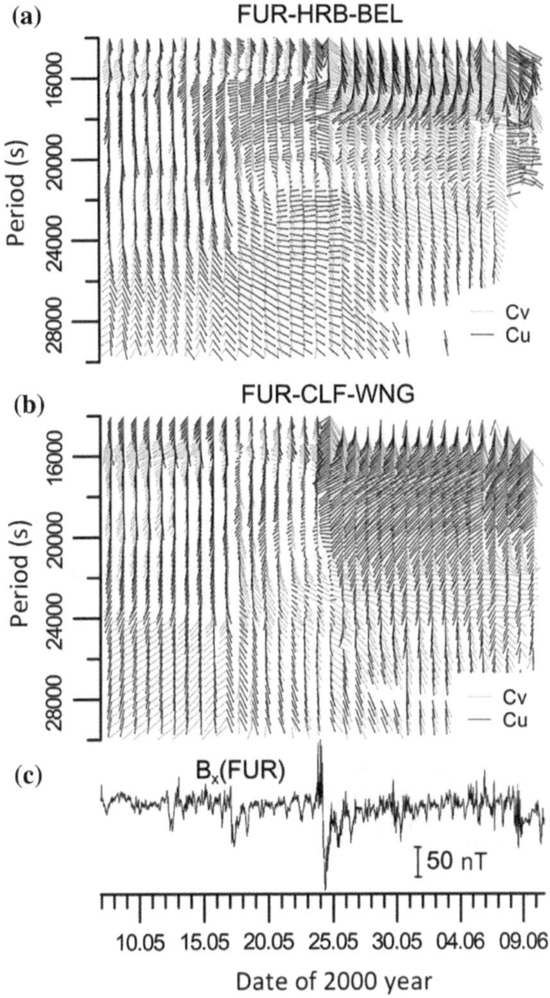

Fig. 5.1 Spectral-time variations of induction arrows in European region: **a** for group of magnetic observatories FUR, HRB & BEL; **b** the same for FUR, CLF, WNG; **c** time series of Bx-component. After Petrishchev and Semenov (2011)

coherences between the input signals B_υ and B_φ should not exceed 0.4 for excluding time intervals with the quasi linear polarization of the tangential field for the GMV method. An example of such analysis is presented in Fig. 5.2.

The results satisfying the above criteria can be grouped for each segment, the median value of which can be taken as the mean one. The median absolute deviation (MAD) can be considered to be a confidence limit estimation.

Transformation of the basic data to such a geomagnetic reference system can be carried out using the polar diagrams of coherences (see 5.3) between tangential field components for each observatory separately (Semenov et al. 2011).

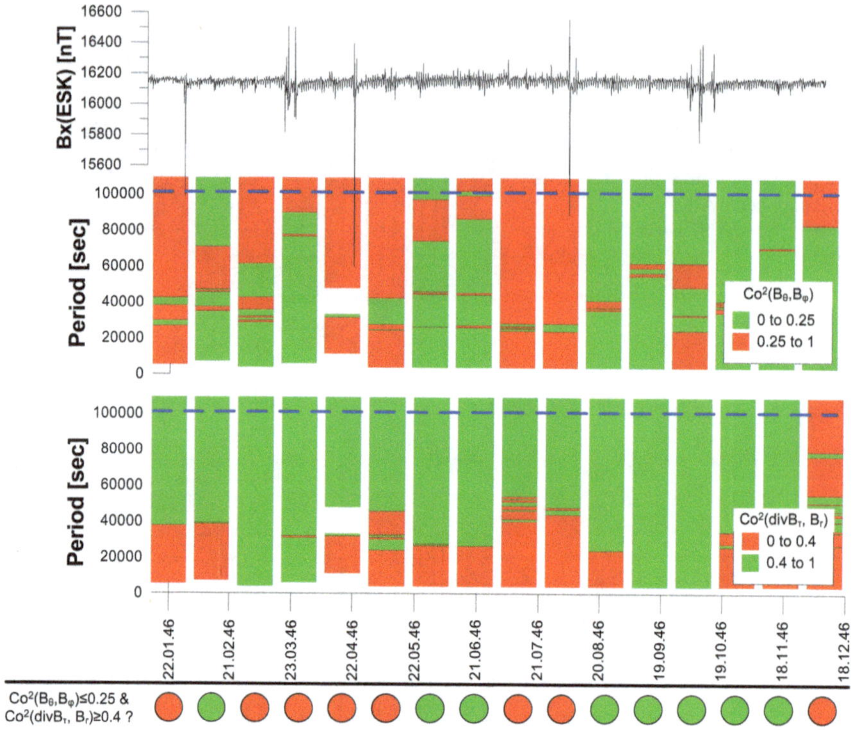

Fig. 5.2 The latitudinal component field at the ESK Observatory (*upper panel*) and the squared coherence for the period of 9 h (*second and third panels*) for processing data from magnetic observatories ESK, CLF and NGK for the year 1946. The fulfillment of coherence criteria is presented on the *bottom line* (*green* is OK, *red* otherwise)

5.3 Variability of the Symmetry Axis of Magnetosphere Ring Current

The geomagnetic pole is formed by the intersection of the symmetry axis of the outer ring current with the Earth's surface. In the method for obtaining the direction to these poles (south or north), only the spectra of two horizontal orthogonal components of the observed magnetic field of the Earth are used (Semenov 1985). This method is very simple, if the source field is a ring current itself, which is the case in practice for periods of field variations over several days.

Indeed, the ring current field held by the dipole magnetic field is considered to be a constant, without strong magnetic storms, and its axis rotates with the Earth's geomagnetic pole (Maus and Lühr 2005). The tangential component of this field must be linearly polarized and focused on the geomagnetic pole. It was shown (Kharin and Semenov 1986) that such a model is valid in general.

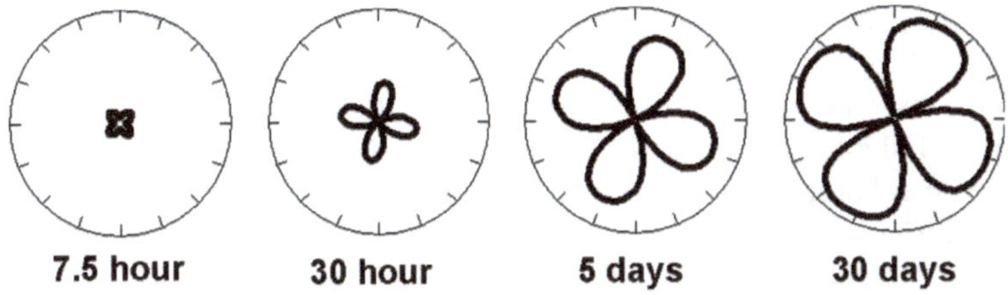

Fig. 5.3 Polar diagrams of the coherence between the components Bx and By in KIV Observatory at several periods of soundings (Semenov et al. 2011)

To determine the direction of the tangential component to the pole, the azimuth of minimum coherence between the orthogonal horizontal components of their polar diagram is used. The maximum value of coherences depends on the power of the ring current field in the total magnetic field in this frequency range. The power of magnetospheric current is not great as long as the period of oscillation does not exceed a few days (Fig. 5.3).

There are two directions of minimal coherence: the first one indicates the direction of a full horizontal field of the outer magnetospheric ring current (the internal field is not large) (arrows in Fig. 5.4), while the second one (orthogonal)

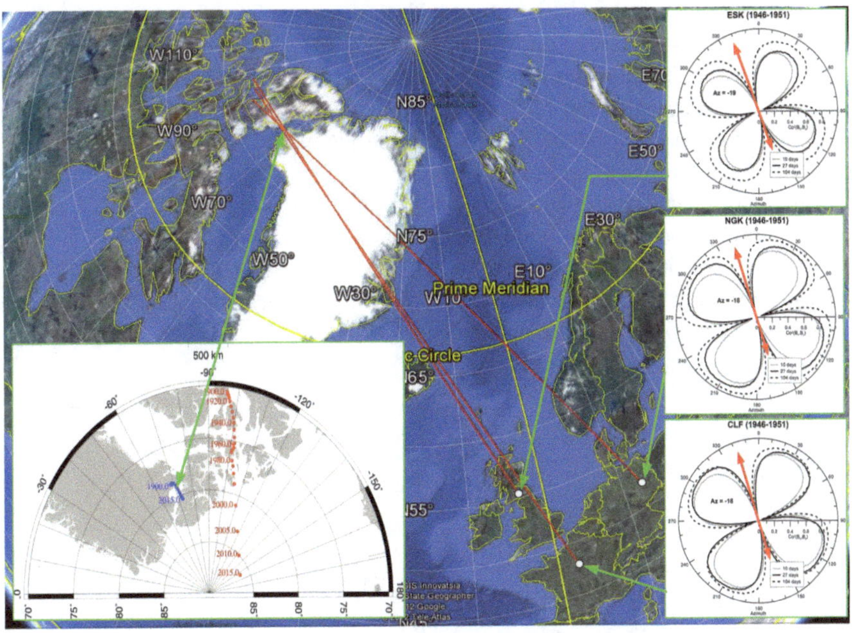

Fig. 5.4 Polar diagram of the coherences between the orthogonal *horizontal* field components for European geomagnetic observatories ESK, NGK and CLF. *Red arrows* show the directions to the geomagnetic poles. Crossing *red lines* indicate the detected pole position and the inset in the lower *left* corner shows the same position from the International Geomagnetic Reference Field model

includes only non-correlated noise and coincides with the direction of the current induced by this field source.

If the observation was made in the geographic coordinate system, then the azimuth of full horizontal field is determined directly from the diagram. To give an example, the hourly values of several observatories in Europe were analyzed (Fig. 5.4). The pole position was estimated by the azimuths of three observatories and the distance between them established from the spherical geometry. The results show that the determination of the position of the geomagnetic poles is better when the observatories that are significantly spaced in longitude are used.

It should be noted that there are two important points: (1) what value of the field is taken as a daily value, and (2) how much suppressed are the diurnal oscillations of the field arising from the rotation of the Earth in an inhomogeneous field with a source in the ionosphere. For example, if you delete all the variations with periods of less than one day and the average daily variation takes the value of fixed-hour UT recording for each observatory, you can see that the azimuth to the geomagnetic pole will change in daily intervals (Fig. 5.5) in the period range from 6 to 20 days.

The maximum deviation of the poles of the magnetospheric source field symmetry can reach ±8° in latitude for days without a strong magnetic storm. Then we should expect distortions in geomagnetic coordinates of the observatory on the same latitude, which will lead to the emergence of the diurnal variation of the field at the observation point.

On a laterally homogeneous sphere with a fixed position of the magnetospheric source it is known (Banks 1969) that $Z_i/Z_j = cos\theta_i/cos\theta_j$ and $H_i/H_j = sin\theta_i/sin\theta_j$. Here the indices i and j refer to the two observatories spaced by co-latitudes (θ), Z being the vertical, and H the full horizontal magnetic field component. Suppose, for example, that the observatory is located in the geomagnetic co-latitude of 45°. Its geomagnetic coordinates associated with the precession of the axis of symmetry of the ring current will vary from 53° to 37° of the same longitude, i.e., generally at 16° per day. In accordance with these formulas and known full field of the ring current (10^6 A), it generates the uniform southward-oriented magnetic field with intensity of about 24 nT parallel to the axis of the Earth's magnetic dipole on the resistive sphere, i.e., in the geomagnetic co-ordinates.

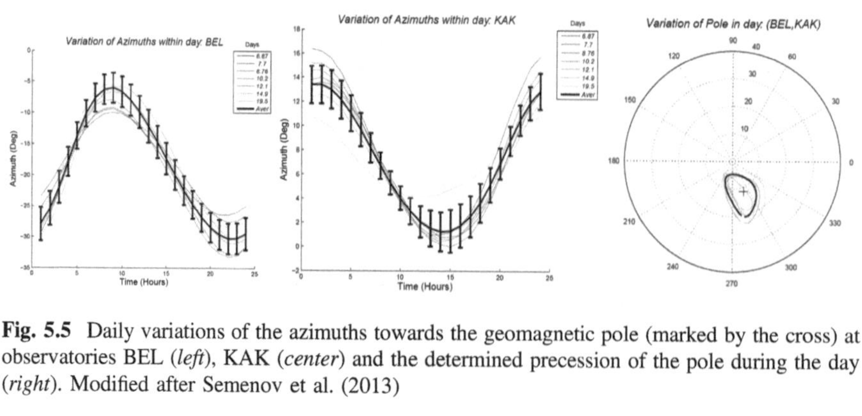

Fig. 5.5 Daily variations of the azimuths towards the geomagnetic pole (marked by the cross) at observatories BEL (*left*), KAK (*center*) and the determined precession of the pole during the day (*right*). Modified after Semenov et al. (2013)

5.4 Relation to the Sun

The GMV method and its variety, the GDS one, were applied to investigate variations of the apparent resistivities on the spherical Earth including sediments, crust and mantle in the middle latitudes of Europe.

The data with the longest synchronous time series were chosen for analysis from the European geomagnetic observatories due to their dense network. The old data measured in magnetic reference system (H, D, Z) have been converted to the geographic one (X, Y, Z). In fact, measurements are carried out and their results are analyzed in a spherical coordinate system (υ, φ, r), where υ is measured from the pole (co-latitude). So, the spherical reference system was used to apply both sounding methods. However, the geographic reference system was used for the GMV method, while the geomagnetic one (θ, λ, r)—for the GDS method. The tangential field divergence is not the measured signal and it was estimated for each observatory group in the same way as in Semenov et al. (2011). The scattered locations of the response functions are shown (Fig. 5.6) in the corresponding triangles formed by the groups of three observatories.

The data analysis was carried out for the 1024-hour interval shifted then to 512 h for the next analysis, and so on. Thus, 17 sounding results per year have been recruited to calculate their yearly values. The time interval of the daily data included 512 days (1.4 year) shifted then to one year (365 or 366 days). Two shorter periods were analyzed by GMV method and two longer by the GDS one.

The synchronous hourly data series for each observatory group have been analyzed by GMV method for the periods of 8.8 h and 1.2 days. The daily data have been analyzed by GDS for periods of 10 and 30 days. The obtained apparent resistivity variations for the observatory group ESK-CLF/VLJ-NGK with the

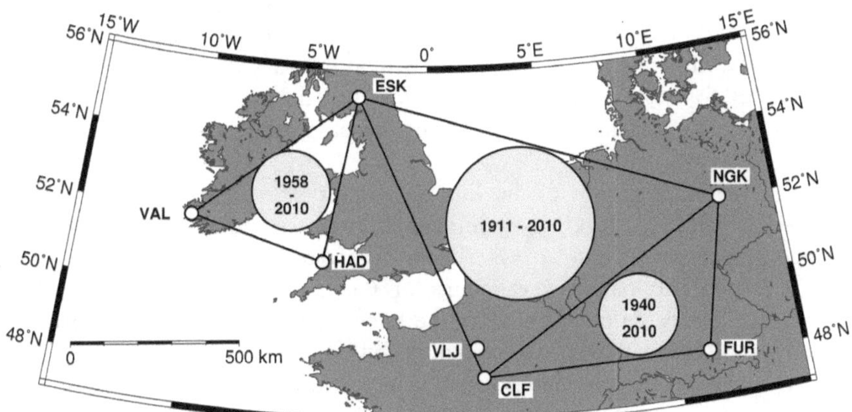

Fig. 5.6 The European geomagnetic observatories with international codes and the sounding areas for groups of three observatories. The analyzed time intervals are dated in years. After Petrishchev and Semenov (2013)

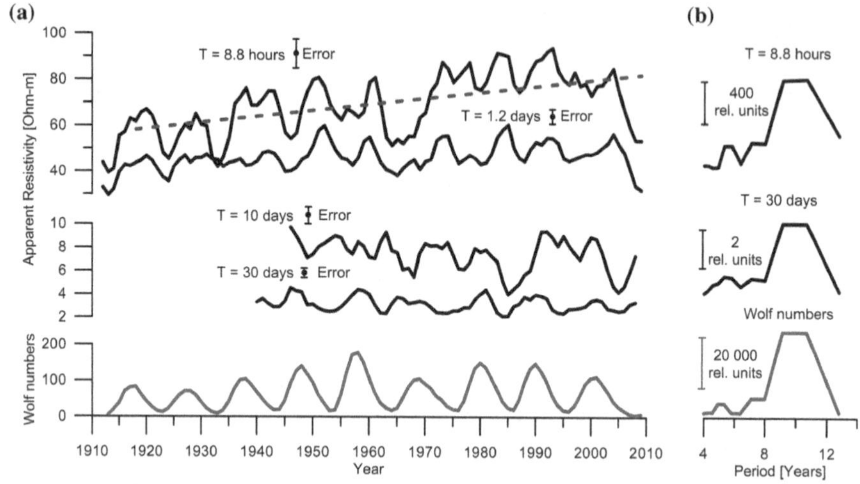

Fig. 5.7 The Earth's apparent resistivity variations detected for the 100-year series of ESK-(VLJ/CLF)-NGK data and the Wolf numbers (**a**); spectral densities of the Wolf numbers and variations with shortest (8.8 h) and longest (30 days) periods in relative units (**b**). After Petrishchev and Semenov (2013)

longest time series are presented in Fig. 5.7 (left) as a smoothed running average of every three years. The quality of data before the years 1940–1950 was insufficient for the analysis of daily data and they are omitted in Fig. 5.7 (left). The spectral analysis of the apparent resistivities has detected pronounced peaks in their spectra around 11 years (Fig. 5.7, right).

The cross spectral analysis of the Earth's apparent resistivities and the Wolf numbers (the yearly numbers of the sunspots) has shown the high coherence between them ($\approx$0.85) and some phase lags of 1–4 years for periods of 6–12 years (GMV method) that are practically absent for the GDS method. Taking into account that the sampling interval was one year, the final accuracy of the phase shift is not less than $\pm$0.5 year. For comparison, the maximal diffusion time to reach a skin depth (see below) for the sounding periods is measured in hours and reaches tens hours.

Then, two other groups of the European observatories, VAL-ESK-HAD and CLF-NGK-FUR (Fig. 5.6), have been chosen to assess a spatial variability of the *C*-responses and induction arrows. These observatories had shorter data series (50–70 years). The obtained variations of the apparent resistivities for these observatory groups definitely confirm the effect detected at the first observatory group by good correlations between all of them (Fig. 4.5 in Sect. 4.3). For example, the correlation between variations of the apparent resistivities of three observatory groups is clearly seen at the results obtained by the GDS method. Note that, roughly speaking, the apparent resistivities are decreasing for the shorter periods, in contrast to those estimated by the GDS method for the last years (2005–2009).

Impedance, as well as C-response, at a fixed period is a functional of conductivity σ (Berdichevsky and Zhdanov 1984), characterizing the conductance over a skin depth. The skin depth, $\mathbf{\Delta} = (\omega\mu\sigma/2)^{-1/2}$, where μ is the magnetic permeability, was introduced theoretically as a penetration depth of induction currents in a homogeneous and conductive half-space (without layers of high resistance) with a plane horizontal boundary (Parkinson 1983). The wavelength in the vertical direction is $\lambda = 2\pi\mathbf{\Delta}$ and its time to reach a skin depth is $\mathbf{\Delta}^2\mu\sigma/2$ (Parkinson 1983). However, the real skin depth for deep soundings must be found for inhomogeneous media, on a sphere, with the resistive Earth's crust where the magnetic permeability can essentially increase with temperature (Kiss et al. 2005). That is why an effective skin depth $\mathbf{\Delta}_{ef} = \mathbf{\Delta}/1.7$ is considered for these media in deep soundings. Another approach has been suggested by Schmucker (1970): $\mathbf{\Delta}'_{ef} = \mathrm{Re}C$. Calculations show that $\mathbf{\Delta}'_{ef} \approx \mathbf{\Delta}_{ef}$ (up to 10%) for depths less than $R_e/10$ as well as for conductances for the plane and spherical Earth.

Changes of the average apparent conductance $S^* = (1/|\rho^*|)\cdot\mathbf{\Delta}_{ef}$, measured in kilo-Siemens (kS), have been estimated for the years 1918 and 2008 for the period of 8.8 h. The apparent resistivity at both skin depths, including the crust and upper mantle, has been increasing from 58 Ω m ($\mathbf{\Delta}_{ef} \approx 430$ km) to 82 Ω m ($\mathbf{\Delta}_{ef} \approx 510$ km), that corresponds to 16% decrease of conductance for 90 years. The same estimates for other periods are comparable with the error bars. So the attempt to explain linear trends of the apparent resistivities (Fig. 5.7) in the frame of the traditional approach has shown implausibly large changes of the Earth's conductance in the upper mantle. Several geophysical processes have similar trends as, for example, the growing distance of the Earth from the mass center of the Solar system, amounting to 10^6 km within a century, or the general increase of the Sun activity in the last century. We will not speculate over the possible reasons.

A feature of the results is the detection of 11-year apparent resistivity variations by both methods, GMV and GDS. These variations have periods that are many times longer than the sounding ones. Note that the C-responses were introduced to exclude influence of source power changes with frequencies equal to the ones registered for the induction sounding on the Earth's surface. An uncorrelated part of the observed field has been mainly suppressed by the robust, coherence-based statistical data processing. The 11-year apparent resistivity variations caused by the outer source are less than 0.2 Ω m (Semenov and Jozwiak 1999), while the detected variations with a period of 8.8 h are of about 10 Ω m or more. An idea was proposed by Boulanger (see Khain and Khalilov 2008): the 11-year resistivity variations can be due to the solar source power, changes of which produce variations of heat in the Earth's conductive layers like the asthenosphere, or mid-mantle layer, as well as their conductivity due to increasing intensity of the induced currents there. This effect would be characterized by phase lags between the solar activity and the mantle conductivity changes at 5–7 years that is caused by the thermal inertia of the mantle rocks (see Khain and Khalilov 2008). However, the phase lags observed by the GMV method are about twice less.

So, secular variations of the Earth's apparent resistivity have been detected by two types of the induction magnetovariation soundings (GMV and GDS) in the period ranges from 8 h to 30 days. The monitoring of hourly and daily data series (50–100 years) was carried out for the geomagnetic observatory data in Europe using the robust and coherency-based statistical processing to obtain the reliable estimates. The existence of linear trend in the apparent resistivities of 8.8 h period has been established.

The 11-year apparent resistivity variations registered by soundings with period $T \approx 9$ h have amplitudes of about 15 Ohm-m, while the same quantity estimated from soundings with T = 11 year is $\sim$0.15 Ohm-m. So the variations from first sounding could be caused by another source. For example, an increase of the solar power can enhance the heating in the high-conductivity mid-mantle layer. Its conductivity will also increase, maybe suddenly, due to larger intensity of the induced currents, which is in agreement with our data. But this effect must be characterized by a phase lag of at least 5–7 years between the solar activity and the mantle conductivity changes, which is caused by the thermal inertia of the mantle rocks (Khain and Khalilov 2008). However, the phase lags estimated by the GDS method are about zero. The source of the variations might be also 11-year variations in the magnetospheric source connected with the Sun activity and determined recently from the satellite data (Lühr and Maus 2010). However, we used only their approximation for deep soundings: some «effective ring current» flowing commonly in a plane of geomagnetic equator. Modeling the sphere with inhomogeneous surface layer shows that apparent resistivity of the GDS sounding depends on a tilt of that plane: the stronger the tilt, the greater the contrast of heterogeneity. A consequence should be a change in the geomagnetic pole position with a period of 11 years.

5.5 Relation to Seismicity

The next point of interest is a relation of induction soundings to seismicity. We will use the same methods—GMV and GDS—for further studies.

Geomagnetic observatories at mid-latitudes of Eurasia ($50° \pm \approx 10°$ of the northern hemisphere) operating at least in the years 1957-2010 have been chosen for deep induction soundings along a profile going from Europe through Asia up to the Pacific Ocean. Locations of these geomagnetic observatories are shown on the background of tectonic structures (Fig. 5.8). Their hourly data for the last 50 years have been analyzed in 17 groups of three neighboring observatories by two deep magnetovariation methods mentioned above. The apparent resistivities have been estimated consecutively for each year generally since 1957 up to 2010 for the period range from 4 h to several months. In Siberia, distances between observatories reach hundreds of kilometers. Such distances required using the spherical reference system; otherwise, the sounding results might have essential scattering. The centers of soundings are shown by triangles in Fig. 5.8, while the radii of soundings can reach many hundreds of kilometers.

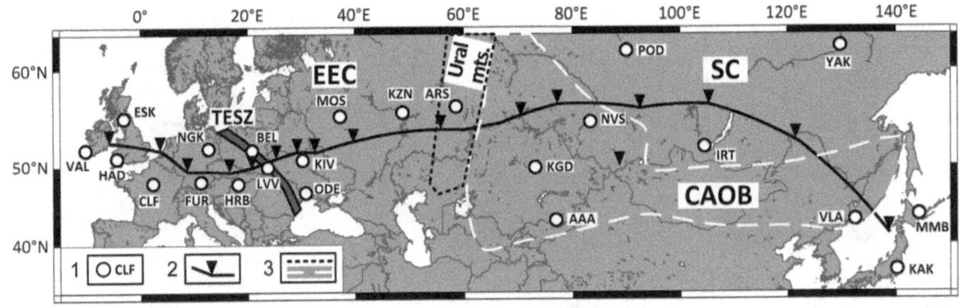

Fig. 5.8 Locations of geomagnetic observatories marked by international codes *1* and profile with centers of soundings. *2* Boundaries of main tectonic structures. *3 TESZ* Trans European Suture Zone; *EEC* East European Craton, *SC* Siberian Craton; and *CAOB* Central Asian Organic Belt

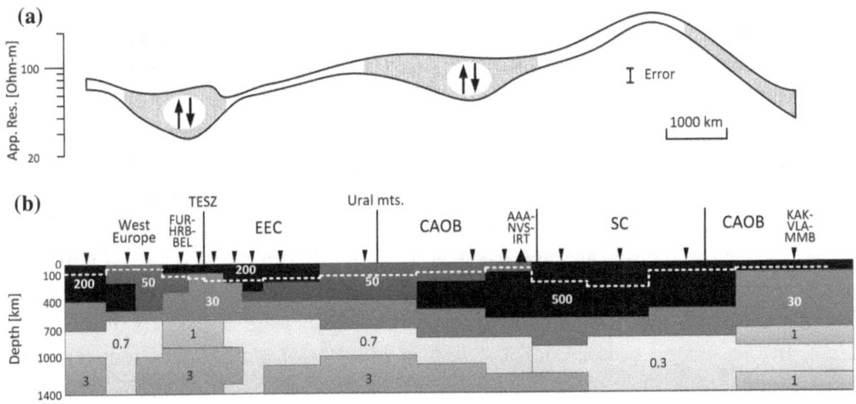

Fig. 5.9 Dispersions of secular apparent resistivities with period T = 9 h (**a**) above weak zones in Eurasia. The geoelectrical model (**b**) is built for 1985–1990; the white dashed line shows the thermal thickness of the lithosphere

The inversion of the obtained responses was performed using the MT2DInv Matlab program (Lee et al. 2009). The result of inversion is presented in Fig. 5.9b. Generally speaking, this resistivity distribution correlates with the thermal thickness of the lithosphere where the temperature of 1300 °C is exceeded (Artemieva 2006).

Three zones with high dispersion, where the change in apparent resistivity for the analyzed period 1957–2010 exceeds the amount of random error, sometimes by an order of magnitude, have been detected along the profile. Such zones are marked in Fig. 5.9a as shaded areas. Their positions are generally consistent with the Trans European Suture Zone in Europe and CAOB in Asia which are characterized by active tectonic processes and, as a consequence, a large number of tectonically weakened zones. Such tectonic formations are characterized by lower values of apparent resistance of the crust than the craton.

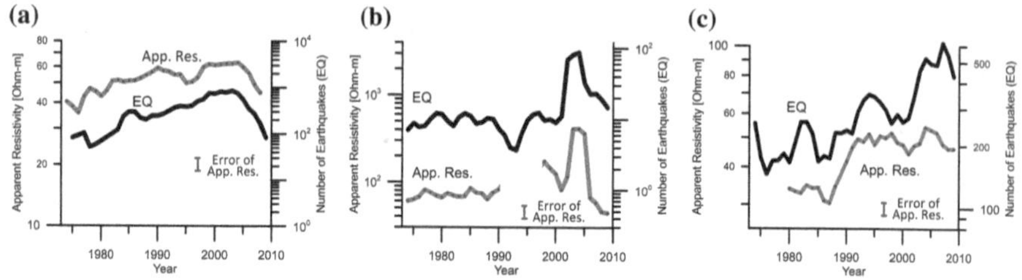

Fig. 5.10 Correlations of secular apparent resistivities (T = 9 h) and EQs in 1973–2010 in: **a** Europe: FUR-HRB-BEL; **b** Asia: AAA-IRT-NVS; **c** Pacific coast: KAK-MMB-VLA

Good correlations between apparent resistivity and change in the number of earthquakes taken from NEIC PDE catalog within a 700 km radius around the center of sounding were established in the first zone (Fig. 5.10a, Europe). Sample earthquakes with magnitudes over 3 were taken from the catalog NEIC PDE.

The second zone has been fixed in Siberia (Fig. 5.10b) where deep soundings have been made for group of stations AAA, IRT and NVS, because exactly there a good earthquake statistics (ten and more events per year) was observed in the years 1973–2010. Such a statistics of earthquakes is absent in Western Siberia. The third zone includes part of the Pacific Ocean shore (Fig. 5.10c), that stands in the way of correct deep soundings.

All the reported cases are characterized by a good correlation between variations in the apparent resistivity and the number of earthquakes within a radius of sensing. Interestingly, most earthquakes in Europe (over 90%) are registered in the upper crust. In the second case (Asia), the largest number of earthquakes focuses registered in the depth range of 25–40 km. Abnormally sharp increase in the number of earthquakes (Fig. 5.10b) was registered at a depth of 10 km for a short time, which is also reflected in the change of apparent resistivity.

Zones where the dispersion is in the same order of magnitude as errors show a lower correlation to seismicity (Fig. 5.11a–b). And zones where the significant dispersion has not been detected do not show a correlation to seismicity (Fig. 5.11c).

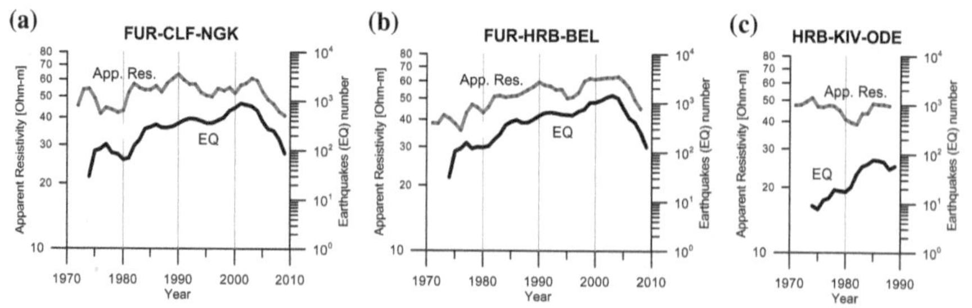

Fig. 5.11 Variation of apparent resistivity for a 9-hour period of sounding and seismicity in 700-km radius from the center of sounding for FUR-CLF-NGK **a** FUR-HRB-BEL **b** and HRB-KIV-ODE **c** groups of observatories. After Semenov and Petrishchev (2012)

A correlation of the number of earthquakes with solar activity was established long ago by Boulanger (see Khain and Khalilov 2008). It was detected that the high seismicity consistently appeared with the 11-year solar cycle on the middle latitudes in a seismically active region of the USSR. It was suggested that changes in solar activity cause fluctuations of the angular velocity of the Earth's rotation which in turn affects the seismic activity. However, the Earth is a layered sphere and angular velocities of its core and mantle can be different. Later on, the core angular moment has been deduced from geomagnetic secular variation model and then "the predicted variations in the length of a day are close to those actually observed" (Love 2008). It means that a source of the secular variation can be located in the mantle. Moreover, the secular variation anomalies could be explained by the laterally inhomogeneous stresses in the Earth's crust based on the piezo-magnetic effect (Sumitomo 1981). Perhaps the same effects were observed for the apparent resistivities synchronously with the lunar–solar tides caused by the oscillation stresses in the crust at the Baltic shield (Zhamaletdinov et al. 2000). One of the physical effects related to pressure is known as "baro-electric" (Grigoriev et al. 1990) which may be considered as a kind of the piezo-electric effect in homogeneously strained conductors. In any case, the changing of stresses plays a significant role in variations of apparent resistivity.

The high correlation between changes of the Earth's apparent resistivities (2–10 times) synchronously with EQ was observed during the recent 30 years. This effect has been ascertained in three zones of Europe and Asia. These regions differed by the depth of earthquakes foci, which has not affected the correlations obtained. In the first zone (Europe), the earthquakes were mostly grouped in the upper crust, in the second zone (Siberia)—in the lower crust, and in the third zone (Pacific Ocean Coast) the earthquakes were distributed in a complicated manner down to a depth of 400 km because of subduction zones.

Correlations between variation of EQ and electromagnetic field give grounds for suggesting the action of the effects of piezoelectric and seismo-electro kinetic phenomena (Guglielmi 2008) in weakened zones and during stresses caused by seismic waves (Jarosinski 2012) from rather powerful earthquakes (Neishtadt et al. 2006; Kurtz and Nibbled 1978). Probably they are a result of seismo-electro-kinetic effects in the sedimentary rocks caused by tectonic stresses (Neishtadt et al. 2006). The first one is the variation of the magnetic field due to elastic stresses, while the second is the variations of media conductivity due to dynamics of fluids in the Earth's crust caused by its stress strain state. The latter effect is characterized by inertia (Svetov 2007). Both effects cannot be described by the Maxwell induction laws only. So a task of future investigations is to separate the induction effect from other kinds of phenomena. This effect is clearly seen in Asia due to specific changes of the earthquakes. Finally, this means that a nature of this effect cannot be purely induction. It can be caused by seismic energy in suture zones between the tectonic plates or near rift zones: the so-called seismoelectric or seismo-magnetic effects.

5.6 Relation to Geomagnetic Jerks

The next point of interest is a relation of induction soundings to geomagnetic jerks. The jerk is conceived as a sudden change in the slope of the secular variation, i.e., the first time derivative of the Earth's magnetic field (De Michelis et al. 2005). Jerks can be detected most clearly in the magnetic declination or the eastern component of the magnetic field. Jerks are clearly seen in the European and Australian data (Love 2008). On the other hand, jerks are not so obvious in the Japanese data. As to the Alaskan data, they show a jerk, but it is of opposite sign to that for Europe and Australia.

Clearly, a global description of the secular variation is complicated. Still, geophysicists have made progress in relating jerks and secular variation to decade-scale changes in Earth's rotational rate that arise from exchanges of angular momentum between the core and the mantle. With certain assumptions, core angular momentum can be deduced from geomagnetic secular variation models. Then, assuming that the Earth's total angular momentum is conserved, one can estimate the changes that should have occurred in the mantle angular momentum over the past century or so. Predicted variations in the length of a day are close to those actually observed, and that gives researchers some confidence that their theories are reasonable.

So, the widespread hypothesis of the geomagnetic jerk nature is based on some internal origin (Malin and Hodder 1982; Love 2008). The fact that jerks are most readily observed at European observatories are largely confined to one component of the field, and are abrupt, argues for a local origin, perhaps a magnetic field instability (Bloxham et al. 2002). Arguing against such an origin is the fact that they represent transitions between long intervals of linear secular variation; in other words, jerks are not simply transient perturbations to the secular variation; instead, they delineate intervals of secular acceleration of opposite sign.

Maybe the Earth's liquid outer core begets a plume reaching even the upper mantle. This phenomenon imposes stringent terms on the conductivity of the Earth's mantle (Semenov and Jozwiak 1999).

Understanding the origin of jerks is important, not only because they are a result of interesting dynamical processes in the core and may help determine the conductivity of the mantle, but also for improving time-dependent models of the geomagnetic field and for strictly practical purpose of forecasting its future behavior, for example, in navigation (MacMillan 2011).

It was found (Petrishchev and Semenov 2013) that the geomagnetic jerk can be seen in the variations of induction vectors. Below we will analyze results of eight mantle MV soundings along a profile in the middle latitudes crossing a region with a powerful jerk of 1969 in Europe. The soundings were separated in time: before and after the jerk occurrence.

We have formed a profile going from the British Islands through Trans-European Suture Zone (TESZ) to the East-European Craton (EEC) using 8 groups of geomagnetic observatories chosen above (Fig. 5.12a). Rate of changes of

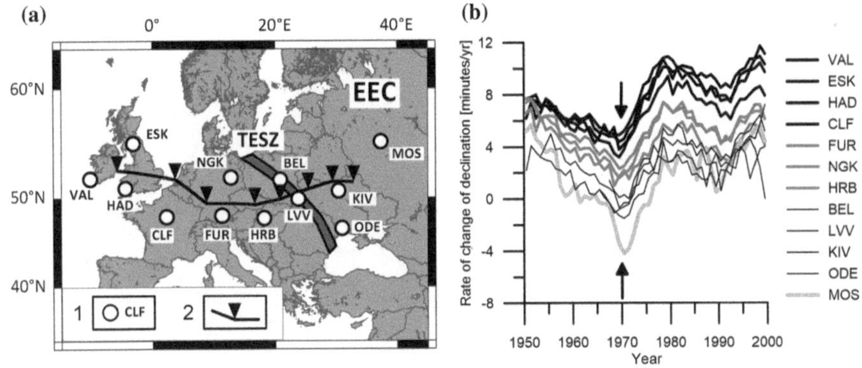

Fig. 5.12 Location of geomagnetic observatories **a** marked by international codes *1* and profile with centers of soundings. *2*. Rate of change of declination on the observatories **b** the geomagnetic jerk of 1969–1970 is highlighted by *arrows*

magnetic declination for each observatory is shown in Fig. 5.12b. Their hourly data for the last 50 years have been processed by the previously mentioned two deep magnetovariation methods and techniques.

The variations of induction arrows $\mathbf{S_u}$ and $\mathbf{S_v}$ for a period of 9 h are presented in Fig. 5.13. It is clearly seen that the induction arrow $\mathbf{S_v}$ is rapidly changing its directions, by about 90°–180°, near the years 1969–1972 for the group of observatories ESK-CLF-NGK and slowly for CLF-NGK-FUR (Fig. 5.13). At the same time, modules of the vector $\mathbf{S_v}$ are reduced by half at both groups. It coincides in time with geomagnetic jerk occurrence on European geomagnetic network. An interesting feature of the magnetic data for those years is the brutal acceleration of the north magnetic pole velocity associated with the 1969 geomagnetic jerk (Mandea and Dormy 2003).

No such essential changes have been detected in induction vectors for 1969–1972 in Eastern Europe (Fig. 5.13e–h), so the main changes in the area considered here were those detected for Western Europe.

Similar significant changes are also established for the apparent resistivity. To test the hypothesis of its connection to jerk, we have fixed mean responses for the group of all the above-mentioned geomagnetic observatories for two 5-year intervals—before and after jerk of 1969. The responses were obtained in the period range of 4 h–2 months. Figure 5.14 shows the responses for ESK-VLJ/CLF-NGK and MOS-KIV-BEL observatory groups in the period range of 4 h–2 days, where the most powerful changes were detected.

For spatial analysis, we have formed a profile from all groups. For obtaining more stable results, the regional response functions up to a period of 11 years (Semenov 1998) have been added to the experimental data and then the inversion MT2DInvMATLAB (Lee et al. 2009) has been applied to the responses. The result of inversion is presented in Fig. 5.15. The difference between the young Phanerozoic European Platform in Western Europe and feebly conductive East

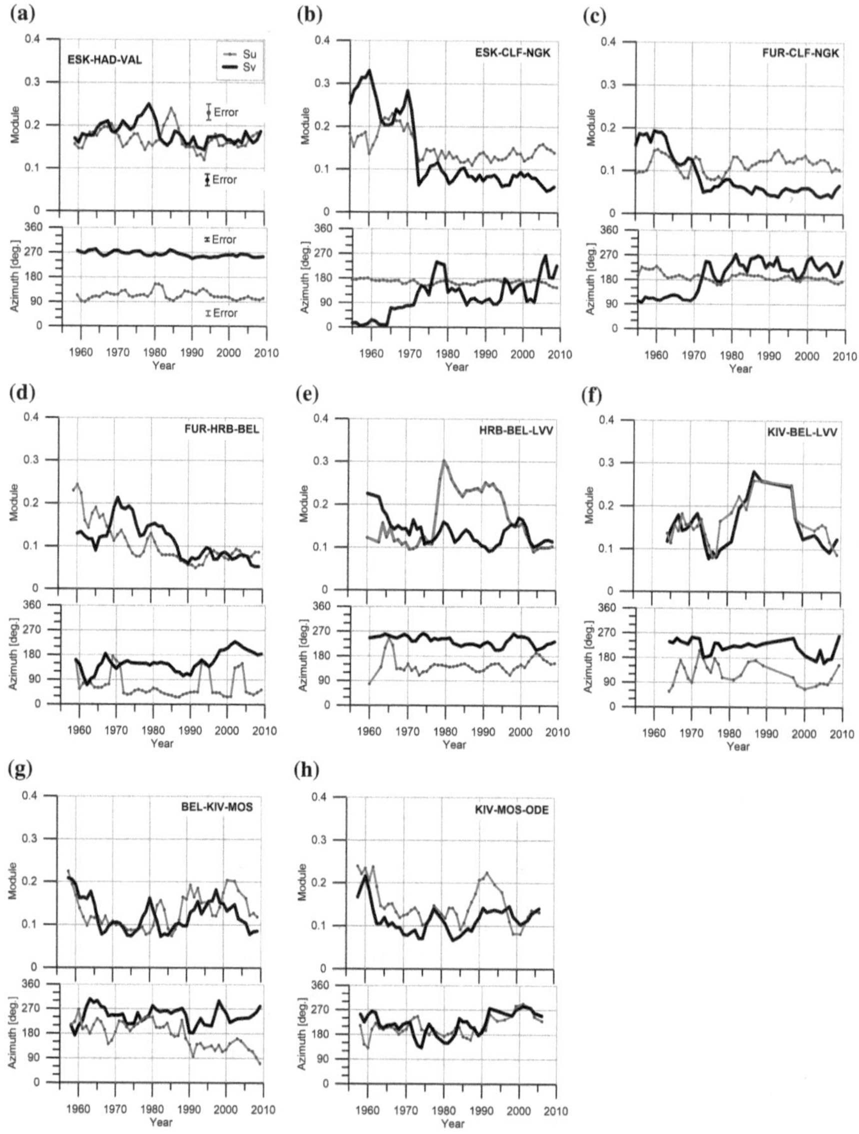

Fig. 5.13 Time variability of induction vectors for period of sounding of 9 h. The observatories for each group are listed on each figure. The legend and error estimations are approximately equal for all groups and are presented in panel *a*

European Craton is clearly seen. The uplifting conductive layer (20–50 Ohm-m) in the upper mantle can be associated with Eifel hotspot (Ritter et al. 2001).

Generally speaking, this resistivity distribution correlated with the thermal thickness of the lithosphere where the temperature of 1300 °C is exceeded (digitized data from Artemieva 2006). It means that deeper than that border the magnetic

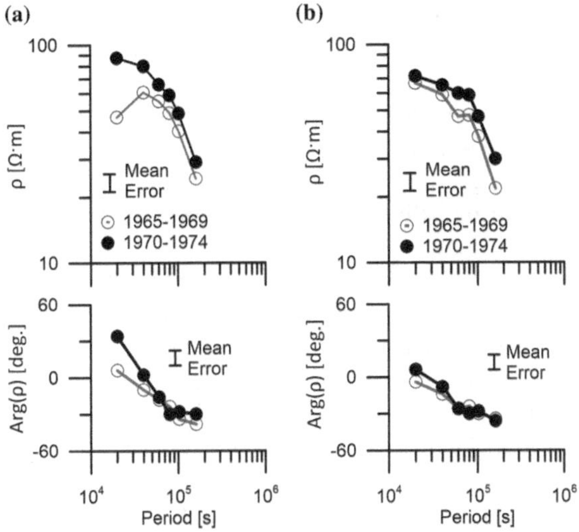

Fig. 5.14 Apparent resistivity (*top*) and its phase (*bottom*) for ESK-VLJ/CLF-NGK **a** and MOS-KIV-BEL **b** groups

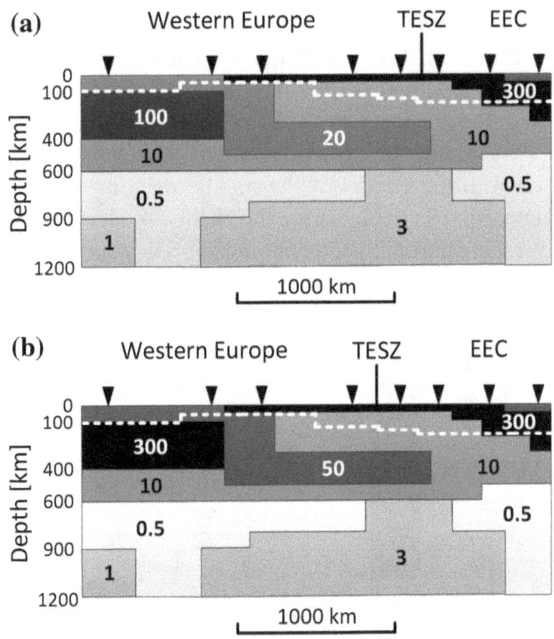

Fig. 5.15 Geoelectrical models through the considered profile according to soundings in the years 1965–1969 **a** and 1970–1974 **b** The *white dashed line* shows the thermal *thickness* of the lithosphere (Artemieva 2006)

properties of matter are changing abruptly; the magnetic permeability can increase many times near this boundary (Kiss et al. 2005) before reaching the Curie-Neel temperature.

The fact that in the upper mantle the resistivity changes coincide with the jerk occurrence has been ascertained. According to the obtained models (Fig. 5.15), the resistivity of the upper mantle increased 2–3 times. It is hard to imagine such fast essential changes (no more than 5 years between the centers of time intervals 1964–1969 and 1970–1974) in so huge, powerful area. So we have decided to check the hypothesis of influence of magnetic permeability in the local layer near the Curie-Neel boundary on the sounding results; such an opportunity is easily following from the apparent resistivity definition. We have used the algorithm for forward modeling that was developed earlier (Petrishchev et al. 2012; Yadav and Lal 1997).

Two 7-layer geoelectrical models have been completed for Western Europe. For the first model, the magnetic permeability equal to the magnetic constant (relative magnetic permeability is equal to 1) was fixed for all layers (Fig. 5.16a). For the second model, we have changed the relative magnetic permeability 2.5 times in the relatively thick layer (Fig. 5.16a) where the Curie-Neel temperature was reached; we used the thermal thickness of the lithosphere from Artemieva (2006). The results of modelling are presented in Fig. 5.16b. It is clearly seen that such minor changes are enough for explaining the effect.

Essential changes in secular apparent resistivity were recently observed while analyzing data of geomagnetic observatories in Europe for the last 50–100 years. Significant rapid changes were detected in induction vectors and response functions in Europe in 1969–1972 that coincide in time with the detection of a geomagnetic jerk. Experimental data have a strong regional feature: the most severe changes were recorded for the ESK-VLJ/CLF-NGK group (Western Europe).

According to experimental data, we have computed two geoelectrical models down to the depths of the middle mantle—before and after a jerk occurrence in time. Models demonstrate the change of electrical resistivity of the upper mantle in Western Europe; the phenomenon has a local character. It is difficult to imagine

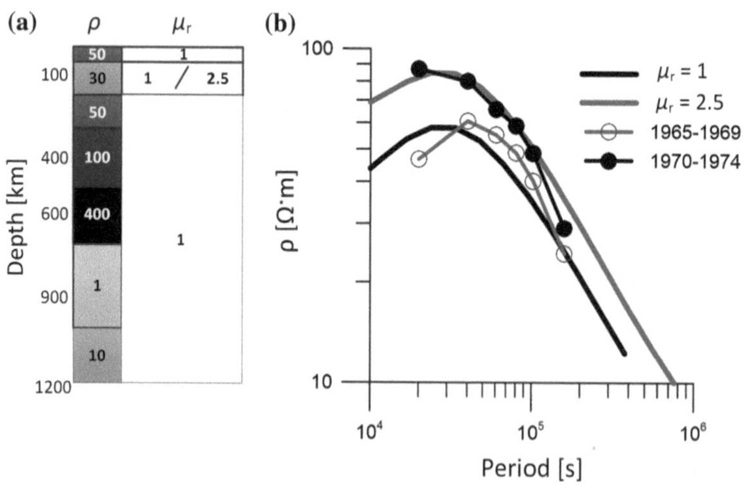

Fig. 5.16 The influence of relative magnetic permeability on the apparent resistivity

such fast changes of the upper mantle conductivity over a large area in such a short time (no more than 5 years). We have tried to describe them by the temperature changes through the magnetic permeability variation near the Curie-Neel temperature (Kiss et al. 2005). The position of this boundary has been taken according to digitized data from Artemieva (2006). We performed modelling by increasing the magnetic permeability in that layer only 2.5 times. That was enough to explain the resulting effect. It is well known that near the boundary where the Curie-Neel temperature is reached it is enough to have relatively small temperature variations in order to get significant changes of the magnetic permeability.

The question as to why these changes occurred precisely in 1969 remains open. Perhaps this may be partly explained by the dynamics of the hydrothermal systems. Each ore-magmatic system is going to be stable as long as possible, and only when all the mechanisms for its maintenance have been exhausted, it jumps to a new steady state characterized by new physical-chemical and geochemical parameters. The main condition of long-life of such systems is a supply from an external source of energy and matter. Such a source or conductor in the study area may be the Eifel hotspot (Ritter et al. 2001), which is directly connected to the upper mantle.

5.7 Conclusion

Thus, in this chapter we have considered the theoretical background of electromagnetic monitoring by deep magnetovariation methods, highlighted the peculiarities of the data processing and revealed a possible connection with solar activity, seismicity and an interesting phenomenon of geomagnetic jerk.

According to the results of monitoring with magnetovariation soundings on the periods from a few hours to several months, the variations of apparent resistivity of the Earth at periods correlated with solar activity cycles have been clearly distinguished. In addition, the accuracy of the analysis allows us to highlight the presence of trends in the apparent resistivities variation, which may indicate a change in the integral conductivity of the mantle on the selected time interval.

The monitoring on a large statistical material allowed establishing (see Sect. 5.5) a number of zones with special properties, where the apparent resistivity variations exceed the amount of random error. An interesting fact is that a strong correlation with the integral seismicity has been detected in these zones.

Correlations between variation of EQ and electromagnetic field give grounds for suggesting the action of piezoelectric and seismo-electro kinetic phenomena (Guglielmi 2008) in weakened zones and during stresses caused by seismic waves (Jarosinski 2012) from relatively powerful earthquakes (Neishtadt et al. 2006; Kurtz and Nibbled 1978).

The most striking phenomenon is a change of the induction vector Sv and apparent change of the Earth's structure established in Western Europe in the years 1969–1972. This phenomenon coincides with the registration of the geomagnetic jerk in the European network of geomagnetic observatories. In this chapter, we have shown (see Sect. 5.6) that the effect can be explained by a change in temperature in the bottom of the thermal lithosphere and, as a consequence, a small change in the relative magnetic permeability.

In any case, the observed effects are rather interesting for further studies, mainly as concerns their formation mechanisms in order to gain new knowledge about the structure and variability of Earth's interior at depths around the core-mantle boundary.

References

Artemieva, I.M.: Global 10 x10 thermal model TC1 for the continental lithosphere: implications for lithosphere secular evolution. Tectonophysics **416**, 245–277 (2006)

Banks, R.J.: Geomagnetic variations and the electrical conductivity of the upper mantle. Geophys. J. R. Astron. Soc. **17**, 457–487 (1969)

Berdichevsky, M.N., Zhdanov, M.S.: Advanced theory of deep geomagnetic sounding, p. 408. Elsevier, Amsterdam (1984)

Bloxham, J., Zatman, S., Dumberry, M.: The origin of geomagnetic jerk. Nature **420**, 65–68 (2002)

De Michelis, P., Tozzi, R., Meloni, A.: Geomagnetic jerks: observation and theoretical modeling. Mem. S. A. It. **76**, 957–960 (2005)

Duma, G., Ruzhin, Y.: Diurnal changes of earthquake activity and geomagnetic Sq-variations. Nat. Hazards Earth Syst. Sci. **3**, 171–177 (2003)

Ernst, T., Jankowski, J., Rozłucki, C., Teisseyre, R.: Analysis of the electromagnetic field recorded in the Friuli seismic zone, northeast Italy. Tectonophysics **224**, 141–148 (1993)

Guglielmi, A.V.: Ultra-low-frequency electromagnetic waves in the Earth's crust and magnetosphere. UFN **177**(12), 1257–1276 (2008)

Guglielmi, A.V., Levshenko, V.T.: Electromagnetic signals from earthquakes. Izv. Phys. Solid Earth **40**(5), 65–70 (1994) (in Russian)

Grigoriev, V.N., Grigorieva, E.V., Postocky, V.S.: Baroelectric effect and magnetic fields of planet and stars. Izv. Phys. Solid Earth **24**(4), 3–14 (1990) (in Russian)

Jarosinski, M.: Compressive deformations and stress propagation in intra continental lithosphere: finite element modeling along the Dinarides—East European Craton profile. Tectonophysics **526–529**, 24–41 (2012). doi:10.1016/j.tecto.2011.07.014

Khain, V.E., Khalilov, E.N.: About Possible Influence of Solar Activity Upon Seismic and Volcanic Activities: Long-term Forecast. Science Without Borders. Transactions of the International Academy of Science H & E., Vol. 3. SWB, Innsbruck (2008). ISBN 978-9952-451-01-6 ISSN 2070-0334

Kharin, E.P., Semenov, VYu.: The curve of deep magnetovariation sounding in Pacific Ocean obtained by continuum spectrum method. France. Ann. Geophys. B. **4**, 329–334 (1986)

Kiss, J., Szarka, L., Pracser, E.: Second-order magnetic phase transition in the Earth. Geophys. Res. Let. **32**, L24310 (2005)

Kurtz, R.D., Niblet, E.R.: Time dependence of magnetotelluric fields in a tectonically active region in Eastern Canada. J. Geomag. Geoelectr. **30**, 561–577 (1978)

Lee, S.K., Kim, H.J., Song, Y., Lee, C.K.: MT2DInvMatlab-A program in MATLAB and FORTRAN for two-dimensional magnetotelluric inversion. Comput. Geosci. **35**(8), 1722–1734 (2009)

Love, J.J.: Magnetic monitoring of earth and space. Phys. Today **61**(2), 31–37 (2008)

Lühr, H., Maus, S.: Solar cycle dependence of quiet-time magnetospheric currents and a model of their near-earth magnetic fields. Earth Planet. Space **62**(10), 843–848 (2010)

Lu, J., Qian, F., Zhao, Y.: Sensitivity analysis of the Schlumberger monitoring array: application to changes of resistivity prior to the 1976 earthquake in Tangshan, China. Tectonophysics **307**, 397–405 (1999)

MacMillan, S.: Encyclopedia of Solid Earth Geophysics: Geomagnetic Field, global pattern/. In: Harsh, K. Gupta (eds.) pp. 373–379. ISBN: 978-90-481-8701-0 (Print) 978-90-481-8702-7 (Online) (2011)

Malin, S.R.C., Hodder, B.M.: Was the geomagnetic jerk of internal or external origin? Nature **296**, 726–728 (1982)

Mandea, M., Dormy, E.: Asymmetric behavior of magnetic dip poles. Earth Planet. Space **55** (153–157), 2003 (2003)

Maus, S., Lühr, H.: Signature of the quiet-time magnetospheric magnetic field and its electromagnetic induction in the rotating earth. Geophys. J. Int. **162**, 755–763 (2005)

Neishtadt, N., Eppelbaum, L., Levitski, A.: Application of seismo-electric phenomena in exploration geophysics: review of Russian and Israeli experience. Geophysics **71**(2), B41–B53 (2006)

Park, S.K., Johnston, M.J.S., Madden, T.R., Morgan, F.D., Morrison, H.F.: Electromagnetic precursors to earthquakes in the ULF band: a review of observation and mechanisms. Rev. Geophys. **31**(2), 117–143 (1993)

Parkinson, W.D.: Introduction to Geomagnetism. Scottish Academic Press, Scotland (1983)

Petrishchev, M.S., Semenov, VYu.: New data about variability of the deep magnetovariation sounding results. Geodynamics **2**(11), 238–240 (2011). (Ukraine)

Petrishchev, M.S., Semenov, VYu.: Secular variations of the Earth's apparent resistivity. Earth Planet. Sci. Lett. **361**, 1–6 (2013). doi:10.1016/j.epsl.2012.11.027

Petrishchev, M.S., Semenov, V.S., Petrova, A.A.: Possible influence of magnetic permeability on the results of induction soundings. Extended Abstracts, 21st EM Induction Workshop, Darwin, Australia, S2-P3, 4 p (2012)

Ritter, J.R.R., Jordan, M., Christensen, U.R., Achauer, U.: A mantle plume below the Eifel volcanic field. Germany. Earth Planet. Sci. Lett. **186**(1), 7–14 (2001). doi:10.1016/S0012-821X(01)00226-6

Saraev, A.K., Pertel M.I., Malkin Z.M.: Monitoring of tidal variations of apparent resistivity. Geologica Acta **8**(1), 5–13 (2010). doi:10.1344/.105.000001512

Schmucker, U.: Anomalies of geomagnetic variations in the south-western United States. Bull. Scripps Inst. Ocean. **13**, 1–165 (1970)

Semenov, V.Y.: Data processing of magnetotelluric soundings. Obrabotka Dannyh Magnetotelluric Soundings, 133 p. (in Russian) (1985)

Semenov, V.Y.: Regional conductivity structures of the earth mantle. Publications of the Institute of Geophysics, Polish Academy of Sciences. Monographic, vol. 65(302) (1998)

Semenov, V.Y., Jozwiak, W.: Model of the geoelectrical structure of the mid- and lower mantle in the Europe-Asia region. Geophys. J. I. **138**, 549–552 (1999)

Semenov, V.Y., Ladanivskyy, B.T., Nowozynski, K.: New induction sounding tested in Central Europe. Acta Geophys. **59**(5), 815–832 (2011)

Semenov, V.Y., Petrishchev, M.S.: Variability of averaged annual impedances of the earth and their spatial gradients in Europe. Geophys. J. **34**(4), 245–252 (2012) (in Russian)

Semenov, V.Y., Hvoždara, M., Vozar, J.: Modeling of deep magnetovariation soundings on the rotating earth. Acta Geophys. **61**, 264–280 (2013). doi:10.2478/s11600-012-0086-9

Sumitomo, N.: Geomagnetic secular variations anomalies in relation to the recent crustal movement in the south-western region of Japan. Bull. Disac. Prev. Res. Inst. **30** (Part 4, No. 274), 97–130 (1981)

Svetov, B.S.: Basics Geoelectrics, p. 647. URRS Publishing, Moscow (2007)

Svetov, B.S., Gubatenko, V.P.: Electromagnetic mechano-electric fields in porous and wet rocks. Izv. Phys. Solid Earth **33**(11), 122–123 (1999) (in Russian)

Tzanis, A.: An examination of the possibility of earthquake triggering by ionosphere-lithosphere electro-mechanical coupling. Hellenic J. Geosci. **45**, 307–316 (2010)

Vanyan, L.L., Kuznetsov, V.A., Lyubetskaya, T.V., Palshin, N.A., Korja, T., Lahti, I., BEAR, W. G.: Electrical conductivity of the crust beneath Central Lapland. Izv. Phys. Solid Earth **38**(10), 798–815 (2002)

Yadav, G.S., Lal, T.: A Fortran 77 program for computing magnetotelluric response over a stratified earth with changing magnetic permeability. Comput. Geosci. **23**(10), 1035–1038 (1997)

Zhamaletdinov, A.A., Mitrofanov, F.P., Tokarev, A.D., Shevtsov, A.N.: The influence of lunar and solar tidal deformations on electrical conductivity and fluid regime of the Earth's crust. Dokl. Earth Sci. **371**(2), 403–407 (2000)

FSC
www.fsc.org
MIX
Papier | Fördert gute Waldnutzung
FSC® C083411

Kolibri 360 GmbH
Ferdinand-Jühlke-Straße 7
99095 Erfurt, Deutschland
produktsicherheit@kolibri360.de